The Art of Algorithm Design

The Art of Algorithm Design

Sachi Nandan Mohanty
Pabitra Kumar Tripathy
Suneeta Satpathy

CRC Press is an imprint of the
Taylor & Francis Group, an **informa** business
A CHAPMAN & HALL BOOK

First edition published 2022
by CRC Press
6000 Broken Sound Parkway NW, Suite 300, Boca Raton, FL 33487-2742

and by CRC Press
2 Park Square, Milton Park, Abingdon, Oxon, OX14 4RN

CRC Press is an imprint of Taylor & Francis Group, LLC

ISBN: 978-0-367-55511-5 (hbk)
ISBN: 978-0-367-55526-9 (pbk)
ISBN: 978-1-003-09388-6 (ebk)

DOI: 10.1201/9781003093886

Typeset in Minion
by codeMantra

This book is dedicated to my Late Father-in-law,
Sukanta Kumar Mohanty - Sachi Nandan Mohanty

This book is dedicated to my parents - Pabitra Kumar Tripathy

This book is dedicated to my parents - Suneeta Satpathy

Contents

Preface

AFTER GAINING A VAST experience in the field of teaching, we decided to share our knowledge and the simplest way of teaching methodologies with the learners residing worldwide. This book is designed to provide fundamental knowledge about the theme of the algorithms and their implementations. No doubt in market a number of books related to this field are available, but the specialty of the current book lies in the emphasis on practical implementations of the respective algorithms through discussion of a number of problems along with its theoretical concepts, so as to enhance the skill of the learners.

This book is designed to provide an all-inclusive introduction to the modern study of computer algorithms along with its proper presentation and substantial intensity keeping in mind that *it* would be handy to all levels of readers. The area of expertise behind this book will familiarize simple concepts like recursion, array, stack, queue, link list, and tree to create an easy understanding of the abstraction of data structure. So, the depth of coverage or mathematical rigor is maintained along with the elementary explanations. Furthermore, each chapter of this book is designed with keen detailed presentation of an algorithm, its specific design technique, application areas and related topics to enforce simple understanding among the learners. Few algorithms that are primarily having theoretical interest are also provided with practical alternatives. Moreover, algorithms are designed using "pseudo-codes" to make it more interesting and comfortable for the learners. Each pseudo-code designed for respective algorithm leads to its clarity and briefness, which can further be translated in any programming language as a straightforward task. As we all know that visualization plays a major role in understanding complex concepts, this book also incorporates detailed working procedures of algorithms by using simple graphical images. All the algorithms in this book are presented with a careful analysis of running times so as to maintain efficiency as a major design criterion.

The text material of this book is primarily intended to be used in undergraduate and graduate courses in algorithm design and data structure. However, the inclusion of engineering issues in algorithms design along with the mathematical apprehensions would make it more comfortable for self-study and well suitable for technical professionals too. Therefore, we look forward to provide the learners with an enjoyable introduction to the field of algorithms with the text material, its step-by-step description, and careful mathematical explanations to eradicate unfamiliarity or difficulty in the subject matter. Learners who have some familiarity with the topic are expected to find the organized chapters with soar preliminary sections proceeding towards more advanced materials.

This book comprises ten separate chapters. Chapter 1 puts forth the preliminary ideas about the concepts of data structures. Chapter 2 narrates the concepts of designing the algorithms and computing the complexities of algorithm as well as the idea about the recurrence relations. Chapter 3 describes the problems related to divide and conquer techniques with practical examples. Chapter 4 discusses the concepts and algorithms related to dynamic programming with appropriate examples. Chapter 5 covers the concepts and algorithms related to greedy choice with apposite examples. Chapter 6 centers on the concepts of different graph algorithms. Chapter 7 emphasizes the approximation algorithm with suitable examples. Chapter 8 deliberates the concepts of matrix operations, linear programming problems and polynomial classes. Chapter 9 highlights different theoretic algorithms. To conclude with, Chapter 10 demonstrates the programming implementations of all the cited algorithms using C-Language.

Authors

Dr. Sachi Nandan Mohanty received his postdoctoral from IIT Kanpur in 2019 and Ph.D. from IIT Kharagpur in 2015, with an MHRD scholarship from the Govt. of India. He has recently joined as an Associate Professor in the Department of Computer Science & Engineering at ICFAI Foundation for Higher Education Hyderabad. Prof. Mohanty's research areas include Data Mining, Big Data Analysis, Cognitive Science, Fuzzy Decision Making, Brain-Computer Interface and Computational Intelligence. Prof. S. N. Mohanty has received three Best Paper Awards during his Ph.D. at IIT Kharagpur from an International Conference on Computer Science & Information Technology at Beijing, China, and another at the International Conference on Soft Computing Applications organized by IIT Roorkee in 2013. He has published 20 research articles in SCI Journals. As a Fellow on Indian Society Technical Education (ISTE), The Institute of Engineering and Technology (IET), Computer Society of India (CSI), Member of Institute of Engineers and IEEE Computer Society, he is actively involved in the activities of the Professional Bodies/Societies.

He has been bestowed with several awards that include "Best Researcher Award from Biju Patnaik University of Technology in 2019," "Best Thesis Award (first Prize) from Computer Society of India in 2015," and "Outstanding Faculty in Engineering Award" from Dept. of Higher Education, Govt. of Odisha in 2020. He has received International Travel funding from SERB, Dept. of Science and Technology, Govt. of India for chairing a session in international conferences held in the United States in 2020. Currently, he is the reviewer of various journals namely *Journal of Robotics and Autonomous Systems* (Elsevier), *Computational and Structural Biotechnology* (Elsevier), *Artificial Intelligence Review* (Springer) and *Spatial Information Research* (Springer). He has edited books published by Wiley, CRC Press, and Springer Nature.

Mr. Pabitra Kumar Tripathy completed his M.Tech. in Computer Science at Berhampur University, Odisha in 2009. He also completed an M.Sc. in Mathematics at Khallikote Autonomous College Berhampur, Odisha in 2003. He is currently pursuing his Ph.D. in Computer Science and Engineering at Biju Patnaik University of Technology, Rourkela, Odisha. He is working as the Head of Department in the Department of Computer Science and Engineering at Kalam Institute of Technology, Berhampur. He has 15 years of teaching and academic experience. His areas of interest are Computer Graphics, Programming Languages, Algorithms, Theory of Computation, Compiler Design, and Artificial Intelligence. He also has published five international journals and has two patents. He has published five books for graduate students.

Dr. Suneeta Satpathy received her Ph.D. from Utkal University, Bhubaneswar, Odisha in 2015, with a Directorate of Forensic Sciences, MHA scholarship from Govt. of India. She is currently working as an Associate Professor in the Department of Computer Science at Sri Sri University, Cuttack, Odisha, India. Her research interests include Computer Forensics, Cyber Security, Data Fusion, Data Mining, Big Data Analysis, Decision Mining and Machine Learning. In addition to research, she has guided many under- and post-graduate students. She has published papers in many international journals and conferences in repute. She has two Indian patents in her credit. Her professional activities also include roles as editorial board member and/or reviewer of *Journal of Engineering Science, Advancement of Computer Technology and Applications, Robotics and Autonomous Systems* (Elsevier) and *Computational and Structural Biotechnology Journal* (Elsevier). She is also editor of several books on different topics such as Digital Forensics, Internet of Things, Machine Learning and Data Analytics published by leading publishers. She is a member of CSI, ISTE, OITS, ACM, IE and IEEE.

CHAPTER 1

Fundamental Concepts of Data Structure

1.1 ARRAY

Whenever we want to store some values, we have to take the help of a variable, and for this, we have to declare it before its use. For example, if we want to store the details of a student, we have to declare the variables as

```
char name [20], add [30];
int roll, age, regdno;
float total, avg;
        etc...
          for an individual student.
```

If we want to store the details of more than one student, then we have to declare a huge number of variables that would increase the length of the program and create difficulty in its access. Therefore, it is better to declare the variables in a group, i.e., the name variable will be used for more than one student, roll variable will be used for more than one student, etc.

Therefore, to declare the variables of the same kind in a group is known as an Array. The concept of array can be used for storing details of more than one student or other objects.

- **Definition**: The array is a collection of more than one element of the same kind with a single variable name.

DOI: 10.1201/9781003093886-1

- **Types of Arrays:**

The arrays can be further classified into two broad categories:

- One-dimensional (the array with one boundary specification)
- Multidimensional (the array with more than one boundary specification)

- One-Dimensional Array

Declaration:

Syntax:

Storage_class Data type variable_name[bound];

The data type may be one of the data types that we have studied. The variable name is also the same as the normal variable_name, but the bound is the number that will further specify the number of variables that you want to combine into a single unit.

Storage_class is optional. By default, it is auto.

Ex: int roll[15];

In the above example, roll is an array of 15 variables that can store the roll_number of 15 students.

In addition, the individual variables are

roll[0], roll[1], roll[2], roll[3], …, roll[14]

1.1.1 Array Element in Memory

The array elements are stored in consecutive memory locations, i.e. the array elements get allocated memory sequentially.

For Ex: int x[7];

If the x[0] is at the memory address 568, then the entire array can be represented in the memory as

x[0]	X[1]	X[2]	X[3]	X[4]	X[5]	X[6]
568	570	572	574	576	578	580

1.1.2 Initialization

The array is initialized just like other normal variables except that we have to pass a group of elements within a chain bracket separated by commas.

Ex: int x[5]= {24, 23, 5, 67, 897};

In the above statement, x[0] = 24, x[1] = 23, x[2] = 5, x[3] = 67, x[4] = 897.

1.1.3 Retrieving and Storing Some Values from/into the Array

Since the array is a collection of more than one element of the same kind, while performing any task with the array, we have to do that work repeatedly. Therefore, while retrieving or storing the elements from/into an array, we have to use the concept of looping.

```
Ex: Write a Program to Input 10 elements into an
array and display them.
#include<stdio.h>
#include<conio.h>
 int main()
      {
      int x[10],i;
      clrscr();
      printf("\nEnter 10 elements into the array");
              for(i=0 ; i<10; i++)
                 scanf(" %d ",&x[i]);
      printf("\n THE ENTERED ARRAY ELEMENTS ARE :");
              for(i=0 ; i<10; i++)
                 printf(" %4d ",x[i]);
      }
```

```
OUTPUT
   Enter 10 elements into the array
   12
   36
   89
   54
   6
   125
   35
   87
   49
   6
   THE ENTERED ARRAY ELEMENTS ARE  :  12   36  89
   54   6 125  35  87  49    6
```

```
/* PROGRAM TO INSERT, DELETE AND SEARCH OF AN
ELEMENT IN AN ARRAY*/
#include<stdio.h>
int insert(int *,int,int,int);
int delete(int *,int,int);
```

```
void input(int *,int);
void display(int *,int);
int search(int *,int,int);
//method to search an element from an array
int search(int *array,int number,int find)
  {
  int i;
    for(i=1;i<=number;i++)
      {  //compare the number to search with array
elements
        if(find == *(array+i))
          {
           return(i);  //return the index
          }
        else
         if(i==number)  //condition for not found
            return(0); //return 0 for failure
      }
    }
    //method to delete an element from an array
int delete(int *array,int number,int position)
 {
   int temp = position;
   while(temp<=number-1)
     {
         *(array+temp)= *(array+(temp+1));
         temp++;
     }
     number= number-1;
     return(number);
 }
 //method to insert an element into an array at
desired location
int insert(int *array,int number,int position,int
element)
  {
   int temp = number;
     while(temp>=position)
       {
          *(array+(temp+1)) = *(array+temp);
          temp -- ;
```

```
        }
      *(array+position)= element;
      number = number+1;
      return(number);
    }
/* INPUT */
  void input(int *array,int number)
    {
      int i;
           for(i=1;i<=number;i++)
            {
             printf("\nEnter the number");
             scanf("%d",array+i);
            }
      }
/*OUT PUT */
   void display(int *array,int number)
     {
       int i;
            for(i=1;i<=number;i++)
             {
              printf("\nValue at the position %d :
          %d",i,*(array+i));
             }
       }
  /* main */
    int main()
     {
      int number,n;
      int *array;
      int position;
      int element;
      int find;
      char ch='y';
      printf("\nInput the number of element into
list");
      scanf("%d",&number);
      array = (int *)malloc(number * 2);
          input(array,number);
           printf("\nEntered list");
      while(ch=='y' || ch=='Y')
          {
```

```
display(array,number);
 printf("\nENTER 1 for INSERT  2 for DELETE 3 FOR
 SEARCH");
 scanf("%d",&ch);
    if(ch==1)
     {
       printf("\nEnter the position to add the data");
       scanf("%d",&position);
       printf("\nEnter the value");
       scanf("%d",&element);
       number = insert(array,number,position,element);
       display(array,number);
     }
 else
       if(ch==2)
       {
         printf("\nEnter the position to delete the
data");
         scanf("%d",&position);
         number =delete(array,number,position);
         display(array,number);
       }
      else
       {
        printf("\nEnter the number to search");
        scanf("%d",&find);
        n=search(array,number,find);
        if(n!=0)
        printf("THE ENTERED NUMBER IS AT %d
POSITION",n);
        else
        printf("\nTHE NUMBER NOT FOUND");
        }
      printf("\nDO YOU WANT TO CONTINUE[y/n]");
      fflush(stdin);
      scanf("%c",&ch);
      }
   return 0;
}
```

OUTPUT

```
Input the number of element into list5

Enter the number12

Enter the number34

Enter the number54

Enter the number66

Enter the number76

Entered list
Value at the position 1 : 12
Value at the position 2 : 34
Value at the position 3 : 54
Value at the position 4 : 66
Value at the position 5 : 76
ENTER 1 for INSERT 2 for DELETE 3 FOR SEARCH3

Enter the number to search54
THE ENTERED NUMBER IS AT 3 POSITION
DO YOU WANT TO CONTINUE[y/n]y

Value at the position 1 : 12
Value at the position 2 : 34
Value at the position 3 : 54
Value at the position 4 : 66
Value at the position 5 : 76
ENTER 1 for INSERT 2 for DELETE 3 FOR SEARCH2

Enter the position to delete the data4

Value at the position 1 : 12
Value at the position 2 : 34
Value at the position 3 : 54
Value at the position 4 : 76
DO YOU WANT TO CONTINUE[y/n]y

Value at the position 1 : 12
Value at the position 2 : 34
Value at the position 3 : 54
Value at the position 4 : 76
ENTER 1 for INSERT 2 for DELETE 3 FOR SEARCH1

Enter the position to add the data4

Enter the value55

Value at the position 1 : 12
Value at the position 2 : 34
Value at the position 3 : 54
Value at the position 4 : 55
Value at the position 5 : 76
DO YOU WANT TO CONTINUE[y/n]n

--------------------------------
Process exited after 39.51 seconds with return value 110
Press any key to continue . . .
```

```
/* TEST SINGULARITY OF A GIVEN MATRIX */
  /* USING CRAMER'S RULES TO FIND DETERMINANT */
# include<stdio.h>
int i, j;
int mat[10][10];
int mat1[10][10];
void display( int, int);
void input( int, int);
int singular(int, int);
/* Output function */
void display( int row, int col)
{
  for(i = 0; i < row; i++)
          {
              for(j = 0; j < col; j++)
              {
                  printf("%5d", mat[i][j]);
              }
              printf("\n");
          }
}
/* Input function */
void input( int row, int col)
{
              for(i = 0 ; i< row; i++)
              {
                  for(j = 0 ;  j<col; j++)
                  {
                      printf("Enter a Number");
                      scanf("%d", &mat[i][j]);
                  }
          }
}
/* Find Determinant using Cramer's rule */
int singular( int row1, int col1)
{
   int i, j, k, l;
   int sum=0, psum=0, partial=0, nsum=0;
   if(row1 == col1)
   {
       printf("\n Number rows  = Number of cols");
       printf("\n Singular Test is possible\n");
```

```
        if(row1 < 3)
        {
             sum = mat[0][0]*mat[1][1] - mat[0]
             [1]*mat[1][0];
             return(sum);
        }
        else
        {
            for(k = 0; k <row1; k++)
                  for(j = 0; j < row1; j++)
                       mat1[k][j] = mat[k][j];
            for(k = 0; k <row1; k++)
                  for(j = row1; j < (2*row1-1); j++)
                       mat1[k][j] = mat1[k][j-row1];
            for(l = 0; l <row1; l++)
            {
                  partial = 1;
                  for(i = 0; i <row1; i++)
                  {
                       partial *= mat1[i][i+l];
                  }
                  psum += partial;
        }
            for(l = row1-1; l < ( 2*row1 -1); l++)
            {
                  partial = 1;
                  for(i =0; i < row1; i++)
                  {
                       partial *=mat1[i][l-i];
                  }
                  nsum += partial;
            }
        sum = psum - nsum ;
             return(sum);
        }
    }
else
        printf("\n Check about singularity is not
        possible");
    return 0;
}
```

```
int main()
{
    int Det;
    int r,c;
    printf("\n Input the number of rows:");
    scanf("%d", &r);
    printf(" Input the number of cols:");
    scanf("%d", &c);
    input(r, c);
    printf("\n Entered  array is as follows:\n");
    display(r, c);
    Det = singular(r, c);
    printf("\n Determinant is : %d", Det);

    if(Det == 0)
        printf("\n The Above matrix is Singular");

    else
        printf("\n Above matrix is not singular");
}
```

OUTPUT

```
 Input the number of rows:3
 Input the number of cols:3
Enter a Number1
Enter a Number3
Enter a Number5
Enter a Number4
Enter a Number2
Enter a Number1
Enter a Number3
Enter a Number4
Enter a Number3

 Entered  array is as follows:
    1    3    5
    4    2    1
    3    4    3

 Number rows   = Number of cols
 Singular Test is possible

 Determinant is : 25
 Above matrix is not singular
--------------------------------
Process exited after 7.099 seconds with return value 30
Press any key to continue . . .
```

1.2 STACK

Stack is a linear data structure that follows the principle of Last in First Out (LIFO). In other words, we can say that if the LIFO principle is implemented with the array, then that will be called the STACK.

The most commonly implemented operations with the stack are PUSH and POP.

Besides these two, more operations can also be implemented with the STACK such as PEEP and UPDATE.

The PUSH operation is known as the INSERT operation, and the POP operation is known as the DELETE operation. During the PUSH operation, we have to check the condition for OVERFLOW, and during the POP operation, we have to check the condition for UNDERFLOW.

- OVERFLOW

If one tries to insert an element in a filled stack, then that situation is called the OVERFLOW condition.

In general, if one tries to insert an element with a filled data structure, then that is called OVERFLOW.

Condition for OVERFLOW

Top=size –1 (for the STACK starts with 0)

Top=size (for the STACK starts with 1)

- UNDERFLOW

If one tries to delete an element from an empty stack, then that situation is called the UNDERFLOW.

In general, if one tries to DELETE an element from an empty data structure, then that is called OVERFLOW.

Condition for UNDERFLOW

Top=–1 (for the STACK that starts with 0)

Top=0 (for the STACK that starts with 1)

Examples

STACK[5]

PUSH(5)

5				

0 1 2 3 4 top=0

PUSH(25)

5	25			

0 1 2 3 4 top=1

PUSH(53)

5	25	53		

0 1 2 3 4 top=2

PUSH(78)

5	25	53	78	

0 1 2 3 4 top=3

PUSH(99)

5	25	53	78	99

0 1 2 3 4 top=4

PUSH(145)

"OVERFLOW" (top=size −1 Condition for OVERFLOW)

POP

5	25	53	78	

0 1 2 3 4 top=3

POP

5	25	53		

0 1 2 3 4 top=2

POP

5	25			

0 1 2 3 4 top=1

POP

5				

0 1 2 3 4 top=0

POP

0 1 2 3 4 top=−1

POP
"UNDERFLOW" (top= −1 Condition for UNDERFLOW)

1.2.1 Algorithm for Push Operation

PUSH(STACK[SIZE], NO, TOP) [STACK[SIZE] is the Stack]
[NO is the Number to Insert]
[Top is the position of the stack]

STEP 1: IF (TOP=SIZE - 1) THEN :
WRITE : "OVERFLOW"
RETURN
[END OF IF]
STEP 2: TOP :=TOP +1
STACK[TOP] := NO
STEP 3: RETURN

1.2.2 Algorithm for Pop Operation

POP(STACK[SIZE], TOP) [STACK[SIZE] is the Stack]
[Top is the position of the stack]

STEP 1: IF (TOP=-1) THEN :
WRITE : "UNDERFLOW"
RETURN
[END OF IF]

```
STEP 2: WRITE : STACK[TOP]
                TOP := TOP -1
STEP 3: RETURN
```

1.2.3 Algorithm for Traverse Operation

```
TRAVERSE(STACK[SIZE], TOP)      [STACK[SIZE] is the Stack]
                                [Top is the position of the stack]
STEP 1: IF (TOP=-1) THEN :
               WRITE : "STACK IS EMPTY "
               RETURN
               [END OF IF]
STEP 2: SET I :=0
STEP 3: REPEAT FOR I=TOP TO 0 BY -1
               WRITE : STACK[I]
               [END OF LOOP]
STEP 4: RETURN
```

1.2.4 Algorithm for Peep Operation

```
PEEP(STACK[SIZE], NO, TOP)      [STACK[SIZE] is the Stack]
                                [NO is the Number to Search]
                                [Top is the position of the stack]
STEP 1: IF (TOP=- 1) THEN :
        WRITE : "STACK IS EMPTY"
        RETURN
        [END OF IF]
STEP 2: SET I: =0
STEP 3: REPEAT FOR I=TOP TO 0 BY -1
           IF (NO=STACK[I]) THEN:
                 WRITE : "NUMBER IS FOUND AT"
                 WRITE : TOP-I+1
                 WRITE : "POSITION"
              RETURN
           [END OF IF]
           IF I=0 THEN:
                 WRITE  : "NUMBER IS NOT FOUND"
           [END OF IF]
         [END OF LOOP]
STEP 4: RETURN
```

OR

```
PEEP(STACK[SIZE], IN, TOP)  [STACK[SIZE] is the Stack]
                            [IN is the Index Number to Search]
                            [Top is the position of the stack]
STEP 1: IF (TOP –IN +1 < 0) THEN :
                WRITE : "OUT OF BOUND"
                RETURN
        [END OF IF]
STEP 2: WRITE : STACK[TOP-IN+1]
STEP 3: RETURN
```

1.2.5 Algorithm for Update Operation

```
UPDATE(STACK[SIZE], NO, TOP)    [STACK[SIZE] is the Stack]
                                [NO is the Number to Update]
                                [Top is the position of the stack]
STEP 1: IF (TOP=- 1) THEN :
                WRITE : "STACK IS EMPTY"
                RETURN
        [END OF IF]

STEP 2: SET I: =0
STEP 3: REPEAT FOR I=TOP TO 0 BY -1
            IF (NO=STACK[I]) THEN:
                    STACK[I]=NO
                RETURN
            [END OF IF]
            IF I=0 THEN:
                WRITE : "UPDATE SUCCESSFULLY NOT COMPLETED"
            [END OF IF]
          [END OF LOOP]
STEP 4: RETURN
```

Program – 1
Wap to perform the PUSH, POP and TRAVERSE operation with the STACK.

```
#include<stdio.h>
#include<alloc.h>
```

```
static int *s, size, top=-1;

void push(int no)
    {
if(top == size-1)
    printf("\n STACK OVERFLOW");
else
 {
   top = top+1;
   *(s+top) = no;
 }
   }
void pop()
  {
if(top == -1)
    printf("\n STACK UNDERFLOW");
  else
    {
  printf("%d IS DELETED", *(s+top));
   --top;
    }
   }
void traverse()
   {
int i;
if(top == -1)
    printf("\n STACK IS EMPTY");
else
   for(i = top; i>=0;i--)
      printf("%5d",*(s+i));
  }
int main()
 { int opt;
printf("\n Enter the size of the stack");
scanf("%d",&size);
s= (int *)malloc(size * sizeof(int));
while(1)
   {
printf("\n Enter the choice");
printf("\n 1.PUSH  2. POP  3. DISPLAY 0. EXIT");
scanf("%d",&opt);
```

```
    if(opt==1)
         {
  printf("\n Enter the number to insert");
  scanf("%d",&opt);
      push(opt);
        }
else
    if(opt==2)
       pop();
   else
     if(opt==3)
       traverse();
      else
        if(opt==0)
           exit(0);
          else
            printf("\n INVALID CHOICE");
    }
}
```

PROGRAM – 2
/* peep operation of the stack using arrays */

```
# include<stdio.h>
# include<ctype.h>
int top = -1,n;
int *s;
/* Definition of the push function */
void push(int d)
{
        if(top ==(n-1))
               printf("\n OVERFLOW");
        else
        {
               ++top;
               *(s+top) = d;
        }
}
/* Definition of the peep function */
void peep()
{
```

```
int i;
      int p;
      printf("\nENTER THE INDEX TO PEEP");
      scanf("%d",&i);

      if((top-i+1) <0)
      {
             Printf("\n OUT OF BOUND");
      }
      else
      {
             Printf("THE PEEPED ELEMENT IS %d",
             *(s+(top-i+1));
      }
}

/* Definition of the display function */

void display()
{
      int i;
      if(top == -1)
      {
             printf("\n Stack is empty");
      }
      else
      {
             for(i = top; i >= 0; --i)
                    printf("\n %d", *(s+i) );
      }
}
int main()             /* Function main */
{        int  no;
         clrscr();
printf("\nEnter the boundary of the stack");
scanf("%d",&n);
      stack = (int *)malloc(n * 2);
     while(1)
      {
      printf("WHICH OPERATION DO YOU WANT TO
      PERFORM:\n");
             printf(" \n 1. Push  2. PEEP 0. EXIT");
```

```
            scanf("%d",&no);
            if(no==1)
            {
                  printf("\n Input the element to
                  push:");
                  scanf("%d", &no);
                  push(no);
                    printf("\n After inserting ");
                             display();
            }
              else
               if(no==2)
                     {
             peep();
                        display();
        }
     Else
       if(no == 0)
                          exit(0);
           else
                         printf("\n INVALID OPTION");
}
```

PROGRAM – 3
/* update operation of the stack using arrays */

```
# include<stdio.h>
# include<ctype.h>
int top = -1,n;
int flag = 0;
int *stack;
void push(int *, int);
int update(int *);
void display(int *);
/* Definition of the push function */

void push(int *s, int d)
{
       if(top ==n-1)
              flag = 0;
       else
```

```
        {
                flag = 1;
                ++top;
                *(s+top) = d;
        }
}
/* Definition of the update function */
int update(int *s)
{
int i;
      int u;
      printf("\nEnter the index");
      scanf("%d",&i);
      if((top-i+1) <0)
      {
                u = 0;
                flag = 0;
      }
      else
      {
                flag = 1;
                u=*(s+(top-i+1));
                printf("\nENTER THE NUMBER TO UPDATE");
                scanf("%d",s+(top-i+1));
      }
      return (u);
      }
/* Definition of the display function */
void display(int *s)
{
      int i;
      if(top == -1)
      {
                printf("\n Stack is empty");
      }
      else
      {
                for(i = top; i >= 0; --i)
                      printf("\n %d", *(s+i) );
      }
}
```

```
int main()
{
      int  no, q=0;
      char ch;
      int top= -1;
printf("\nEnter the boundary of the stack");
scanf("%d",&n);
      stack = (int *) malloc(n * 2);
    up:
      printf("WHICH OPERATION DO YOU WANT TO
      PERFORM:\n");
            printf(" \n Push->i\n update->p");
            printf("\nInput the choice : ");
            fflush(stdin);
            scanf("%c",&ch);
            printf("Your choice is: %c",ch);
            if(tolower(ch)=='i')
            {
                  printf("\n Input the element to
                  push:");
                  scanf("%d", &no);
                  push(stack, no);
                  if(flag)
                  {
                        printf("\n After inserting ");
                        display(stack);
                        if(top == (n-1))
                               printf("\n Stack is
                               full");
                  }
                  else
                        printf("\n Stack overflow
                        after pushing");
            }
             else
             if(tolower(ch)=='p')
             {
                  no = update(stack);
                  if(flag)
                  {
```

```
                        printf("\n The No %d is
                        updated", no);
                  printf("\n Rest data in stack is as
                  follows:\n");

                        display(stack);
                  }
                  else
                        printf("\n Stack underflow" );
                        }
            opt:
    printf("\nDO YOU WANT TO OPERATE MORE");
    fflush(stdin);
    scanf("%c",&ch);
     if(toupper(ch)=='Y')
      goto up;
        else
          if(tolower(ch)=='n')
            exit();
              else
                {
                  printf("\nINVALID CHARACTER...Try
                  Again");
                   goto opt;
                  }
}
```

1.3 QUEUE

Queue is a linear data structure that follows the principle of FIFO. In other words, we can say that if the FIFO principle is implemented with the array, then that is called the QUEUE.

The most commonly implemented operations with the Queue are INSERT and DELETE.

Besides these two, more operations can also be implemented with the QUEUE such as PEEP and UPDATE.

During the INSERT operation, we have to check the condition for OVERFLOW, and during the DELETE operation, we have to check the condition for UNDERFLOW.

The end at which the insertion operation is performed will be called the REAR end, and the end at which the delete operation is performed is known as the FRONT end.

- **Types of Queue**
  - Linear Queue
  - Circular Queue
  - D – Queue (Double-ended queue)
  - Priority Queue.
- **Linear Queue**
  - OVERFLOW

    If one can try to insert an element with a filled QUEUE, then that situation is called an OVERFLOW condition.
- **Condition for OVERFLOW**

Rear = size –1 (for the QUEUE that starts with 0)
Rear = size (for the QUEUE that starts with 1)

  - UNDERFLOW

    If one can try to delete an element from an empty QUEUE, then that situation is called as UNDERFLOW condition.
- **Condition for UNDERFLOW**

Front = –1 (for the QUEUE that starts with 0)
Front = 0 (for the QUEUE that starts with 1)

- **Condition for EMPTY QUEUE**

Front = –1 and Rear = –1 [for the QUEUE that starts with 0]
Front = 0 and Rear = 0 [for the QUEUE that starts with 1]

Examples

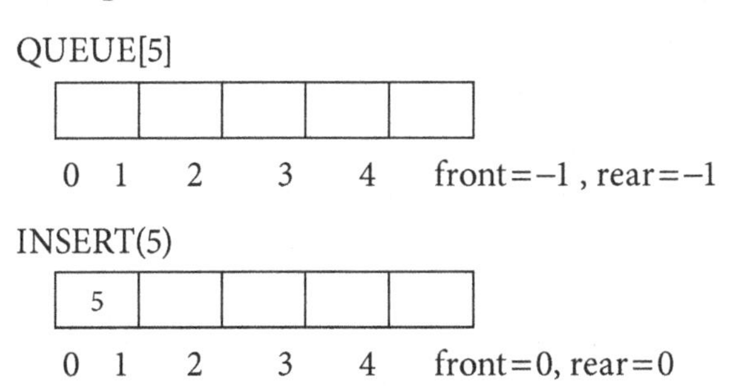

INSERT(25)

5	25			

0 1 2 3 4 front=0, rear=1

INSERT(53)

5	25	53		

0 1 2 3 4 front=0, rear=2

INSERT(78)

5	25	53	78	

0 1 2 3 4 front=0, rear=3

INSERT(99)

5	25	53	78	99

0 1 2 3 4 front=0, rear=4

INSERT(145)

"OVERFLOW" (rear=size -1 Condition for OVERFLOW)

DELETE

	25	53	78	99

0 1 2 3 4 front=1, rear=4

DELETE

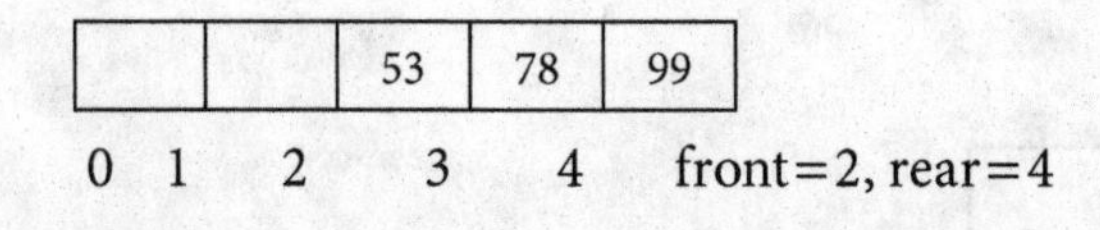

DELETE

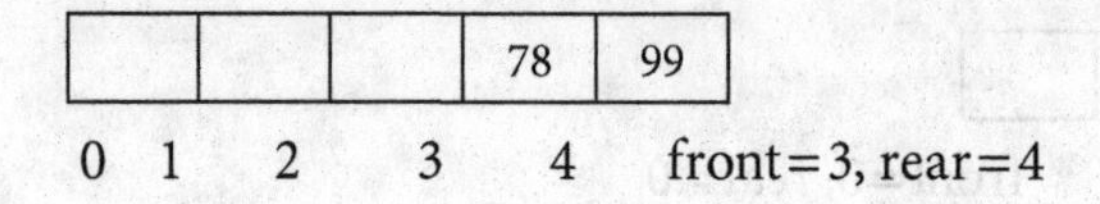

DELETE

				99

0 1 2 3 4 front=4, rear=4

DELETE

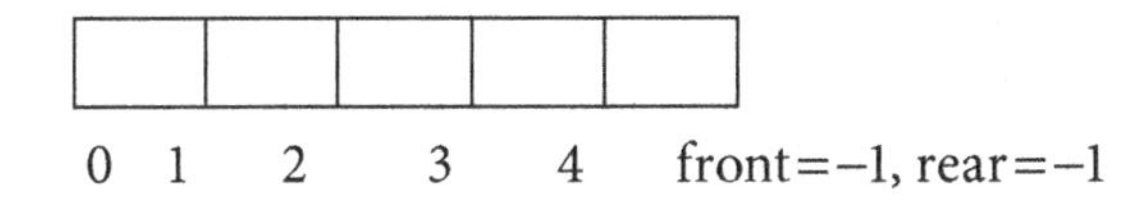

DELETE
"UNDERFLOW" (front=–1 Condition for UNDERFLOW

1.3.1 Algorithm for Insert Operation

INSERT(QUEUE[SIZE], FRONT, REAR, NO)
STEP 1: IF (REAR=SIZE - 1) THEN :
    WRITE : "OVERFLOW"
    RETURN
  [END OF IF]
STEP 2: IF (REAR=-1) THEN :
    FRONT := 0
    REAR :=0
  ELSE :
    REAR :=REAR+1
  [END OF IF]
STEP 3: QUEUE[REAR] :=NO
STEP 4: RETURN

1.3.2 Algorithm for Delete Operation

DELETE(QUEUE[SIZE], FRONT, REAR)
STEP 1: IF (FRONT=-1) THEN :
    WRITE : "UNDERFLOW"
    RETURN
  [END OF IF]
STEP 2: WRITE: QUEUE[FRONT]
STEP 3: IF (FRONT ==REAR) THEN :
    FRONT := -1
    REAR :=-1

```
        ELSE :
            FRONT := FRONT +1
        [END OF IF]
STEP 4: RETURN
```

1.3.3 Algorithm for Traverse Operation

```
TRAVERSE(QUEUE[SIZE], FRONT, REAR)
STEP 1 : IF (FRONT=-1) THEN :
            WRITE: "QUEUE IS EMPTY"
            RETURN
        [END OF IF]
STEP 2 : SET I:=0
STEP 3 : REPEAT FOR I= FRONT TO REAR
            WRITE: QUEUE[I]
        [END OF LOOP]
STEP 4:  RETURN
```

1.3.4 Algorithm for Peep Operation

```
PEEP(QUEUE[SIZE], NO, FRONT, REAR)
                                    [QUEUE[SIZE] is the Stack]
                                    [NO is the Number to Search]
                                    [Front & Rear are the positions of
                                    the stack]
STEP 1: IF (REAR=- 1) THEN :
                WRITE : "STACK IS EMPTY"
                RETURN
        [END OF IF]
STEP 2: SET I: =0
STEP 3: REPEAT FOR I=FRONT TO REAR
            IF (NO=QUEUE[I]) THEN:
                    WRITE : "NUMBER IS FOUND AT"
                    WRITE : I+1
                    WRITE : "POSITION"
                RETURN
        [END OF IF]
        IF I= REAR THEN:
                WRITE  : "NUMBER IS NOT FOUND"
```

    [END OF IF]
  [END OF LOOP]
STEP 4: RETURN

1.3.5 Algorithm for Update Operation

UPDATE(QUEUE[SIZE], NO, FRONT, REAR)
[QUEUE[SIZE] is the QUEUE]
[NO is the Number to Update]
[FRONT & REAR is the position of the stack]

STEP 1: IF (REAR =- 1) THEN :
      WRITE : "STACK IS EMPTY"
      RETURN
  [END OF IF]
STEP 2: SET I: =0
STEP 3: REPEAT FOR I=FRONT TO REAR
    IF (NO=QUEUE[I]) THEN:
      QUEUE[I]=NO
     RETURN
   [END OF IF]
   IF I=REAR THEN:
     WRITE : "UPDATE SUCCESSFULLY NOT COMPLETED"
   [END OF IF]
  [END OF LOOP]
STEP 4: RETURN

PROGRAM – 4
/*INSERTION AND DELETION IN A QUEUE ARRAY IMPLEMENTATION */

```
# include<stdio.h>
int *q,size,front=-1,rear=-1;
void  insert(int n)
{
        if(rear ==size-1)
            printf("\n QUEUE OVERFLOW");
     else
```

```
      {
       rear ++;
              *(q+rear) = n ;
              if(front == -1)
                      front = 0;
       }
}
/* Function to delete an element from queue */
void Delete()
{
      if (front == -1)
      {
              printf("\n Underflow");
              return ;
      }
      printf("\n Element deleted : %d", *(q+front));
      if(front == rear)
      {
              front = -1;
              rear = -1;
      }
      else
              front = front + 1;
}
void display()
{
      int i;
      if (front == -1)
              printf("\n EMPTY QUEUE");
      else
        {
printf("\nTHE QUEUE ELEMENTS ARE");
      for(i = front ; i <= rear; i++)
              printf("%4d", *(q+i));
      }
}
int main()
{
      int opt;
printf("\n Enter the size of the QUEUE");
scanf("%d",&size);
q= (int *) malloc(size * sizeof(int));
```

```
while(1)
   {
printf("\n Enter the choice");
printf("\n 1.INSERT  2. DELETE  3. DISPLAY 0. EXIT");
scanf("%d",&opt);
    if(opt==1)
       {
  printf("\n Enter the number to insert");
  scanf("%d",&opt);
     insert(opt);
       }
else
    if(opt==2)
       Delete();
   else
     if(opt==3)
       display();
      else
        if(opt==0)
           exit(0);
          else
            printf("\n INVALID CHOICE");
    }
}
```

1.4 LINKED LIST

The linked list is the way of representing the data structure that may be linear or nonlinear. The elements in the linked list are allocated memory randomly with a relation in between them. The elements in the linked list are known as NODES.

The link list is quite better than the array due to the proper usage of the memory.

- **Advantage of Link List over the Array**

  The array always requires the memory that is in sequential order, but the linked list requires a single memory allocation that is sufficient to store the data. In the case of an array, the memory may not be allotted even if the available memory space is greater than the required space because that may not be in sequential order.

- **Types of Link List**

The link list is of four types:

- Single Link List
- Double Link List
- Circular Link List
- Header Link List

1.4.1 Single Link List

1.4.1.1 Structure of the Node of a Linked List

The node of a link list has the capacity to store the data as well as the address of its next node and the data may vary depending on the user's requirement, so it is better to choose the data type of the node as STRUCTURE that would have the ability to store different types of elements. The general format of the node is

```
Struct tagname
  {
Data type member1;
Data type member2;
……………………..
…………………….
…………………...
Data type membern;
Struct tagname *var;
  };
```

Example:

```
struct link
  {
int info;
struct link *next;
  };
```

This structure is also called as self-referential structure.

- **Concept of Creation of a Linked List**

```
int *p,q=5;
    p=&q;
    *p=*p+5
```

After this, the value of q is being changed to 10.

The main observation here is that if a pointer variable points to another variable, then whatever changes that are made with the pointer will directly affect the variable whose address is stored inside the pointer and the concept is used to design/create the linked list.

- **Logic for Creation**

```
    struct link
      {
     int info;
    struct link *next;
      };
struct link start, *node;
```

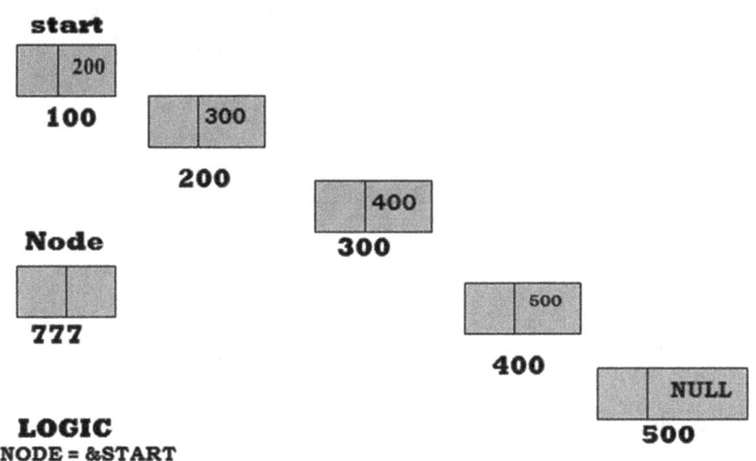

LOGIC

```
NODE = &START
Node->next = (struct link * )malloc(sizeof(struct link))
Node = node->next
Node->next = NULL
Input node->info
```

- **Algorithm for Creation of Single Link List**

```
struct link
 {
 int info;
struct link *next;
 };
CREATE(START,NODE)    [START IS THE STRUCTURE TYPE OF VARIABLE]
                      [NODE IS THE STRUCTURE TYPE OF POINTER]
STEP-1 : NEXT[START]: = NULL
STEP-2 : NODE := ADDRESS OF START
STEP-3 : ALLOCATE A MEMORY TO NEXT[NODE]
         NODE :=NEXT[NODE]
         INPUT : INFO[NODE]
         NEXT[NODE] : = NULL
STEP-4 : REPEAT STEP-3 TO CREATE MORE NODES
STEP-5 : RETURN
```

- **Algorithm for Traversing of Single Link List**

```
struct link
 {
 int info;
struct link *next;
 };
TRAVERSE(START,NODE)[START IS THE STRUCTURE TYPE OF VARIABLE]
                    [NODE IS THE STRUCTURE TYPE OF POINTER]
STEP-1 : NODE := NEXT[START]
STEP-2 : REPEAT WHILE (NEXT[NODE] #NULL )
            WRITE : INFO[NODE]
      Node:=NODE[NEXT]
         END OF LOOP
STEP-3 : RETURN
```

- **Insertion**

The insertion process with link list can be discussed in four different ways:

- Insertion at Beginning
- Insertion at End
- Insertion when node number is known
- Insertion when information is known

- **Programs**

```
/* CREATION OF A LINKED LIST */
#include<stdio.h>
#include<alloc.h>

struct link
{
      int info;
      struct link *next;
};
struct link *start = NULL; /*empty list*/
/* Function main */
int main()
{
      struct link *node;
      link(node);
      display(node);
}
void link(struct link *);
void display (struct link *);

void link(struct link *node)
{
      char ch='y';
      while(ch=='y' || ch=='Y')
  {
            printf("\n Enter a number : ");
            scanf("%d",&node->info);
            node->next=(struct link* ) malloc (sizeof
            (struct link));
            node=node->next;
            node->next=start;
            printf("\nDO YOU WANT TO CREATE FURTHER
            NODES[Y/N]: ");
            ch=getche();
  }
}
void  display(struct link *node)
{
```

```
        printf("\n THE VALUES OF THE ENTERED NODES
        ARE\n");
         do
          {
               printf("%5d",node->info);
               node=node->next;
          }while(node->next!=NULL);
}
```

/* INSERT A NODE INTO A SINGLE LINKED LIST AT DESIRED LOCATION */

```
# include <stdio.h>
# include <alloc.h>
struct link
  {
      int info;
      struct link *next;
  };
int i;

struct link start,*previous,*new;
void insert(struct link *);
void create(struct link *);
void display(struct link *);

int main()
{                            /* Function main */
      struct link *node;
      create(node);
      printf("\n THE LINK LIST IS AS FOLLOWS :\n");
      display(node);
      insert(node);
      printf("\n After Inserting THE LINK LIST IS AS
      FOLLOWS :\n");
      display(node);
}

void create(struct link *node)
{                      /* Function create a linked list */
      char ch='y';
      start.next = NULL;
      node = &start; /* Point to the start of the list */
      i = 0;
```

```
      while(ch =='y' || ch=='Y')
       {
            node->next = (struct link* )
            malloc(sizeof(struct link));
            node = node->next;
            printf("\nENTER A NUMBER");
            scanf("%d", &node->info);
            node->next = NULL;
            fflush(stdin);
            printf("\nDO YOU WANT TO CONTINUE[Y/N] ");
            fflush(stdin);
            scanf("%c",&ch);
            i++;
        }
      printf("\n Total nodes = %d", i);
 }


void insert(struct link *node)
{                           /* Inserting a node */
      int non = 0;
      int pos;
      node = start.next;
      previous = &start;
      printf("\n ENTER THE POSITION TO INSERT:");
      scanf("%d",&pos);
      while(node)
      {
            if((non+1) == pos)
            {
                  new = (struct link* ) malloc
                  (sizeof(struct link));
            new->next = node ;
            previous->next = new;
            printf("\n Input the node value: ");
            scanf("%d", &new->info);
            break ;
      }
      else
      {
            node = node->next;
            previous= previous->next;
```

```
            }
            non++;
        }
}

void display(struct link *node)
{                               /* Display the list */
        node = start.next;

        while (node)
        {
            printf(" %d", node->info);
            node = node->next;
        }
}
```

```
/* DELETING A NODE FROM A SIMPLE LINKED LIST WHEN NODE
NUMBER IS KNOWN */
# include <stdio.h>
# include <alloc.h>
struct link
{
        int info;
        struct link *next;
};
int i;
struct link start, *previous;

void  display(struct link *);
void  delete(struct link *);
void  create(struct link *);
/* Function main */
int main()
{
        struct link *node;
        clrscr();
        create(node);
        printf("\n THE CREATED LINKED LIST IS :\n");
        display(node);
        delete(node);
      printf("\n AFTER DELETING THE NODE THE LINKED LIST
      IS ");
```

```
      display(node);
}

void create(struct link *node)
{
char ch='y';
start.next = NULL;
   node = &start;
while(ch == 'y' || ch=='Y')
      {
            node->next = (struct link* ) malloc
            (sizeof(struct link));
            node = node->next;
            fflush(stdin);
            printf("\n ENTER THE NUMBER");
            scanf("%d",&node->info);
            node->next = NULL;
            printf("\n DO YOU WANT TO CREATE MORE
            NODES");
            fflush(stdin);
            scanf("%c",&ch);
            i++;
      }
      printf("\n THE LENGTH OF LINKED LIST IS = %d", i);
}
void display(struct link *node)
{
      node = start.next;
      printf("\n Created list is as follows:\n");
      while (node)
      {
            printf(" %d", node->info);
            node = node->next;
      }
}
void delete(struct link *node)
{
      int n = 1;
      int pos;
      node = start.next;
      previous = &start;
```

```
        printf("\n Input node number you want to
        delete:");
        scanf(" %d", &pos);
        while(node)
        {
                if(n == pos)
                {
                        previous->next = node->next;
                        free(node);
                        break ;
                }
                else
                {
                        node = node->next;
                        previous = previous->next;
                 }
                n++;
        }
}
```

1.5 TREE

A tree is a nonlinear data structure in which the elements are arranged in the parent and child relationship manner. We can also say that in the tree data structure, the elements are stored in a sorted order. And it is used to represent the hierarchical relationship.

A TREE is a dynamic data structure that represents the hierarchical relationships between individual data items.

In a tree, nodes are organized hierarchically in such a way that

- There is a specially designated node called the root, at the beginning of the structure except when the tree is empty.
- Lines connecting the nodes are called branches, and every node except the root is joined to just one node at the next higher level (parent).
- Nodes that have no children are called leaf nodes or terminal nodes.

```
/*CREATION OF A BINARY SEARCH TREE*/
#include<stdio.h>
# include<alloc.h>
```

```
struct node
{
      int info;
      struct node *left;
      struct node *right;
};

struct node *create(int, struct node *);
void display(struct node *, int);
int main()
{
      int info ;
      char ch='y';
      struct node *tree ;
      tree = NULL;
      while(ch == 'y' || ch=='Y')
      {
            printf("\n Input information of the
            node: ");
            scanf("%d", &info);
            tree = create(info, tree);
            printf("\n Tree is ");
            display(tree, 1);
            printf("\nDO YOU WANT TO CREATE MORE
            CHILD [Y/N]");
            scanf("%c",&ch);
      }
}
struct node * create(int info, struct node *n)
{
      if (n == NULL)
      {
            n = (struct node *) malloc( sizeof(struct
            node ));
            n->info = info;
            n->left = NULL;
            n->right= NULL;
            return (n);
}
            if (n->info >= info )
            n->left = create(info, n->left);
else
```

```
            n->right = create(info, n->right);
      return(n);
}
void  display(struct node *tree, int no)
{
      int i;
      if (tree)
      {
            display(tree->right, no+1);
            printf("\n ");
            for (i = 0; i < no; i++)
                  printf("    ");
            printf("%d", tree->info);
            printf("\n");
            display(tree->left, no+1);
      }
}
```

- **Traversing with Tree**

  The tree traversing is the way to visit all the nodes of the tree in a specific order. The Tree traversal can be accomplished in three different ways:

  - INORDER traversal
  - POSTORDER traversal
  - PREORDER traversal

  Tree traversal can be performed in two different ways:

  - BY USING RECURSION
  - WITHOUT USING RECURSION

- **Recursively**

  - Inorder traversal
  - Traverse the Left Subtree in **INORDER(Left)**
  - Visit the Root node
  - Traverse the Right Subtree in **INORDER(Right)**

- Preorder traversal
  - Visit the Root Node
  - Traverse the Right Subtree in **PREORDER(Left)**
  - Traverse the Right Subtree in **PREORDER(Right)**
- Postorder traversal
  - Traverse the Right Subtree in **POSTORDER(Left)**
  - Traverse the Right Subtree in **POSTORDER(Right)**
  - Visit the Root Node

1.5.1 Questions

1. What is the key benefit of the Linked list over Array?
2. Explain the application of the stack with a suitable example.
3. Apply the concept of queue using a link list.
4. Apply the concept of stack using a linked list.
5. Differentiate between STACK and QUEUE.
6. Write the nonrecursive procedure for traversing a TREE data structure.
7. What is the basic difference between a linear queue and a circular queue?
8. Form a Binary Search Tree by considering 12 months of the year.
9. What is dynamic memory allocation and what is the benefit of static memory allocation?
10. How to represent a tree data structure? Explain in detail.

CHAPTER 2

Concepts of Algorithms and Recurrences

2.1 ALGORITHM

An algorithm can be defined in different ways as given below:

- It is a set of rules for carrying out calculations either by hand or by a machine.
- It is a finite number of step-by-step procedures to achieve a required result.
- It is a sequence of computational steps that transform the input into the output.
- It is a sequence of operations performed on the data, which has to be organized in data structures.
- It is an abstraction of a program to be executed on a physical machine (mode of computation).

An algorithm can also be viewed as a tool for solving a well-specified computational problem.

OR

A finite sequence of unambiguous steps terminating in a finite amount of time for all possible inputs of finite size.

DOI: 10.1201/9781003093886-2

The study of an algorithm can be done in four steps given below:

1. Design of algorithms
2. Algorithm validation
3. Analysis of algorithms
4. Algorithm testing.

2.2 DESIGN OF ALGORITHMS

Several methods are used to design an algorithm, but out of those, some most commonly used algorithms are

- Divide-and-conquer
- Incremental approach
- Dynamic programming
- Brute force approach
- Decrease and conquer
- Backtracking
- Greedy approach.

In general, an algorithm can be designed by undertaking the following three steps.

1. Algorithm Validation

   This step checks the algorithm result for all sets of legal input. It is a highly essential step to check for the validation of an algorithm after writing it. In this step, the algorithm is not converted into a program. After showing the validity of the methods, a program can be written by using any one of the languages. This phase is also known as program proving or program verification.

2. Analysis of Algorithms

   In most cases, for a specific problem, several solutions may be proposed in terms of algorithms, and it's too difficult to choose a proper algorithm that will result in the best output. In this step, analysis of the number of algorithms has to be performed to get the best algorithm, which means there will be a comparison among the algorithms.

The analysis of the algorithm focuses on time and space complexities. The amount of memory required by the program to run to completion is referred to as Space Complexity. In addition, for an algorithm, time complexity depends on the size of the input; thus, it is a function of input size "n." The amount of time needed by an algorithm (program) to run to completion is referred to as Time Complexity.

It should be remembered that different time requirements may arise for the same algorithms. In general, we have to deal with the best-case time, average-case time and worst-case time for an algorithm.

3. Algorithm Testing

   This phase includes the testing of the program. It consists of two phases:

- Debugging phase
- Performance measurement.

  Debugging is the process of finding and correcting errors. Performance measurement describes that the correctness of the program for any set of input data.

2.3 ALGORITHMIC NOTATIONS

No doubt that neither the algorithm has to follow any of the computer languages nor it has to be remembered that for writing the algorithm, we must follow certain rules to maintain uniformity. The algorithms are provided in the form of pseudo-codes, which means the algorithms are language- and machine-independent.

- Each algorithm should be identified by a unique name. The name of the algorithm has to begin with the word "Algorithm" to reflect the name of the algorithm.
- Each algorithm should have a description of its own, which means what are the variables used in it and what is the logic of that.
- The algorithm must have some comments. The comment lines are represented by the symbol ► or [ ]
- Every algorithm should be represented as different steps.
- For assignment, the operator such as "o"→ should be used.

- We can use the common operators such as "<", "<", "≤", "≥", "≠", "=" etc.
- The logical operators such as "and," "or" and "not" can also be used.
- For exchanging the values, we may use "↔"
- Use the word "set" for assigning/exchanging the values.
- Variables can be either local or global. The array variables can also be used exactly as the same as in C language.
- In algorithm, we can also use the conditional structures as
  - If
  - If…else
  - If…else…if
- Like switch case in algorithm, we can use select case as

```
Select case (expression)
{
case constant1:
                ......................
case constant2:
                ......................
default:
                .......................
}
```

- For looping construct, use the format as
  - Repeat for var → sequence
  - Repeat while (condition)
  - Repeat for var → sequence while (condition)
  - Repeat through step for var → sequence

Example

Repeat for I → x, x+1, x+2, … x+n

Or

Repeat for I → x to x+n
Or
Repeat for I → 1 to n (n ≤ 20)
Repeat while (n ≤ 20)
Repeat for I → 1 to n while (n ≤ 20)

- The goto statement may be used to transfer the control to another step.
- The exit loop statement may be used to terminate the loop, and it works for a single loop only so if the nested loop is used, then there we may use it to terminate a single loop.
- The exit statement may be used at the end of the algorithm to terminate the algorithm.

Example

```
main()                      Time          Cost
 {                            1             C1
  int i,j,k,n;                1             C2
  i = i + 1;                n + 1           C3
  for (i = 1; i <= n;i++)     n             C4
    {                         1             C5
     j = j + 1;
    }
  k = k + 1;
 }
```

1. Line number 1st, 2nd, and, 5th are running 1 time only.
2. Line number 3rd running n+1 times in the algorithm. Similarly inside for-loop line number.
3. Line number 4th is running n times in the algorithm.

Therefore, calculating the above example, time complexity with respect to n times term used in algorithm as follows:

=> $T(n) = C1*1 + C2*1 + C3*(n+1) + C4*n + C5*1$
=> $T(n) = 1*(C1 + C2 + C5) + C3*n+C3*1 + C4*n$
=> $T(n) = 1*(C1+C2+C3+C5) + n*(C3 + C4)$
=> $T(n) = O(n)$

As you know time, Complexity is expressed in big O-notation. Here, time complexity of the recent algorithm is O(n).

In this algorithm, the bigger value is n*(C3+C4) where getting the highest time taken is n, i.e., why value is O(n).

4. Don't confuse that Expression was C3*(n+1) so the value should be O(n+1) but it is not. Value will be O(n) because when it is fully simplified, it will give result n*(C3+C4), i.e., only n term value should be used in big O Notation.

Example:

Describe an algorithm for finding the maximum value in a finite sequence of integers.

Solution:

Procedure max(a1, a2, a3, …, an: integers)
max = a1
for i = 2 **to** n
if max < ai, **then** max = ai
output max
Number of steps: $1 + (n - 1) + (n - 1) + 1 = 2n$

Example:

```
char h = 'y'; // This will be executed 1 time
int abc= 0; //This will be executed 1 time
```

For a loop like

```
for (int i = 0; i < N; i++)
{
  Printf('Hello World');
}
```

int i=0; this will be executed only **once**. The time is calculated to i=0 and not the declaration.

i < N; this will be executed **N+1** times

i++ ; this will be executed **N** times

Therefore, the number of operations required by this loop is

{1+(N+1)+N} = 2N+2

Example

```
For i= 1 to n;
  j= 0;
while(j<=n);
  j=j+1;
```

here, the total number of executions for inner loop are n+1, and the total number of executions for outer loop are n(n+1)/2, so the total number of executions for whole algorithm are n+1+n(n+1/2) = (n^2+3n)/2. Here, n^2 is the dominating term, so the time complexity for this algorithm is O(n^2).

Example

```
for ( i = 0; i < N; i++ ) {
  for ( j = 0; j < N; j++ )
    statement;
}
```

is quadratic. The running time of the two loops is proportional to the square of N. When N doubles, the running time increases by N * N.

Example

```
void Main()
     {
          char[] arr = { 'a', 'b', 'b', 'd', 'e' };
          char invalidChar = 'b';
              int ptr = 0, N = strlen(arr);
              for (int i = 0; i < n; i++)
              {
                  if (arr[i] != invalidChar)
                 {
                     arr[ptr] = arr[i];
                     ptr++;
                 }
            }

             for (int i = 0; i < ptr; i++)
             {
```

```
                printf("%d",arr[i]);
            printf(' ');
        }

    }
  }
}
```

Output for the above code will look like

```
a d e
for (int i = 0; i < N; i++)
    {
        if (arr[i] != invalidChar)
        {
            arr[ptr] = arr[i];
            ptr++;
        }
    }
```

The above code snippet contains many basic operations, which will be repeated. The basic operations it contains are

int i=0;	This will be executed only once The time is calculated to i=0 and not the declaration
i<N;	This will be executed N+1 times
i++ ;	This will be executed N times
if(arr[i]!=invalidChar)	This will be executed N times
arr[ptr]=arr[i];	This will be executed N times (in the worst-case scenario)
ptr++;	This will be executed N times (in the worst-case scenario)

Therefore, the number of operations required by this loop is {1+(N+1)+N}+N+N+N = 5N+2

The part inside the curly braces is the time consumed by the Loop alone (i.e., for(int i=0;i<N;i++)), it is 2N+2. Therefore, we should remember that it is usually the same (unless you have a non-default FOR loop)

Now for the second loop

```
for (int i = 0; i < ptr; i++)
{
     printf("%d",arr[i]);
   printf(' ');
}
```

Remember, a loop takes 2N+2 iterations. Therefore, here it will take 2ptr+2 operations. Again, considering the worst-case scenario, ptr will be N, so the above expression evaluates to (again) 2N+2.

Then there are these two additional operations of Console, which will be executed N times each (again, the worst-case scenario).

Therefore, the above code snippet will take

{1+(N+1)+N}+N+N = 4N+2

char[] arr = { 'a', 'b', 'b', 'd', 'e' };	This will be executed N times
char invalidChar = 'b';	This will be executed 1 time
int ptr = 0;	This will be executed 1 time
N = strlen(arr)	This will be executed 1 time
Printf("\n")	This will be executed 1 time

Note: The character array initialization will execute N times. This is because you are assigning one character at a time.

Therefore, the rest of the code requires N+4

Adding everything up I get

(N+4)+(5N+2)+(4N+2) = 10N+8

So the asymptotic time complexity for the above code is O(N), which means that the above algorithm is a linear time complexity algorithm.

Example

```
main()                          Time        Cost
 {                              1           C1
 int i,j,a,b;                   n+1         C2
 for(i=1; i<=n;i++)             n*(m+1)     C3
    {                           n*m         C4
     for(j=1; j<=m;j++)         n           C5
        {
          a=a+1;
        }
     b=b+1
    }
 }
```

1. Line number 1 in the algorithm is running one time.
2. Line number 2 in the algorithm is running n+1 times, i.e., first this 1st line is compiled by the compiler, and second it runs n times more to run 3rd and 4th lines in the algorithm.
3. Line number 3 in the algorithm running n*(m+1) times, i.e., first for loop is run n times, and second it runs n*m times more to run line number 4 in the algorithm.
4. Line number 4 runs by both loops n*m times in the above algorithm.
5. Line number 5 runs n times by line number 2 in the algorithm.

- Calculating the time complexity of the above algorithm of the post as follows:

  => T(n) = C1*1 + C2*(n+1) + C3*(n*(m+1)) + C4*(n*m) + C5*n

  => T(n) = C1*1 + C2*n + C2*1 +C3*(nm + n) + C4*nm +C5*n

  => T(n) = 1*(C1+C2) + n*(C2 + C3 + C5) + nm*(C3 + C4)

  => T(n) = O(nm)

  As you notice in the above expression, the highest value is nm*(C3+C4); therefore, the value of the time complexity of the above algorithm is O(nm).

Example:

What is the worst-case complexity of the following code fragments?

```
for (i = 0; i < N; i++)
{
    sequence of statements
}
for (j = 0; j < M; j++)
{
   sequence of statements
}
```

Solution

The first loop is O(N) and the second loop is O(M).

Since you don't know which is bigger, you say this is O(N+M).

This can also be written as O(max(N,M)).

In the case where the second loop goes to N instead of M, the complexity is O(N). So, O(N+M) becomes O(2N), and when you drop the constant, it is O(N). O(max(N,M)) becomes O(max(N,N)) which is O(N).

Example:

What is the worst-case complexity of the following code fragments?

```
for (i = 0; i < N; i++)
{
   for (j = 0; j < N; j++)
{
      sequence of statements
   }
}
for (k = 0; k < N; k++)
{
   sequence of statements
}
```

Solution

The first set of nested loops is $O(N^2)$, and the second loop is O(N). This is $O(\max(N^2,N))$, which is $O(N^2)$.

Example:

What is the worst-case complexity of the following code fragments?

```
for (i = 0; i < N; i++)
{
   for (j = N; j > i; j--)
{
      sequence of statements
   }
}
```

Solution

This is very similar to our earlier example of the nested loop where the number of iterations of the inner loop depends on the value of the index of the outer loop.

The only difference is that in this example, the inner-loop index is counting down from N to i+1. It is still the case that the inner loop executes N times, then N – 1, then N – 2, etc., so the total number of times the innermost "sequence of statements" executes is $O(N^2)$.

2.3.1 Calculation for Method Calls

When a statement involves a method call, the complexity of the statement includes the complexity of the method call. Assume that you know that method f takes constant time, and that method g takes time proportional to (linear in) the value of its parameter k. Then the statements below have the time complexities indicated.

```
f(1);   // O(1)
g(k);   // O(k)
```

When a loop is involved, the same rule applies. For example:

for (j = 0; j < N; j++) g(N);

has complexity (N^2).

The loop executes N times, and each method call g(N) is complexity O(N).

Example:

```
for (j = 0; j < N; j++) f(j);
```

Solution

Each call to f(j) is O(1). The loop executes N times, so it is $N \times O(1)$ or O(N).

Example:

```
for (j = 0; j < N; j++) g(k);
```

Solution

Each time through the loop g(k) takes k operations and the loop executes N times. Since you don't know the relative size of k and N, the overall complexity is O(N×k).

2.4 GROWTH OF FUNCTION

2.4.1 Asymptotic Notations

These notations are used to describe the running time of an algorithm. This shows the order of growth of function. The symbolic notations are of five types such as

- O – Notation (Big-Oh notation)
- Ω – notation (Big-omega)
- ω – notation (Little-omega)
- θ – notation (Theta)
- o – notation (Little-oh)

2.4.2 O – Notation (Big – Oh Notation)

The Big-Oh notation provides the upper bounds for a specific function.

We can say that f(n) = O(g(n)) iff there exist +ve constants c and n0 such that

$$f(n) \leq c * g(n) \text{ for all } n, n \geq n_0$$

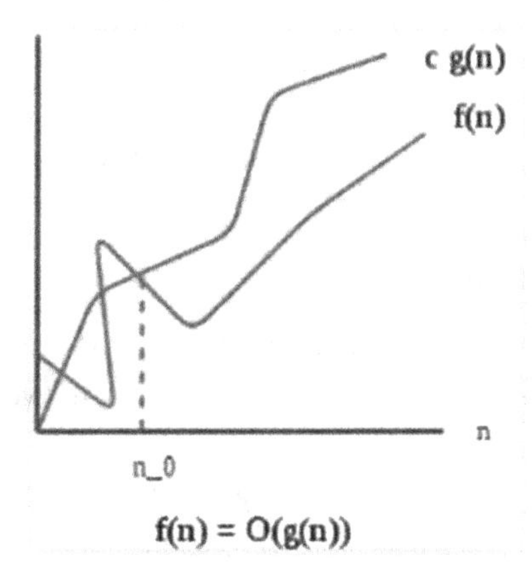

f(n) = O(g(n))

O(1) is the computing time for a constant.
O(n) is for linear.
$O(n^2)$ is called quadratic.
$O(n^3)$ is called cubic.
$O(2^n)$ is called exponential.
O(log n) is faster than O(n) for large value of n.
O(n log n) is better than $O(n^2)$ but not as good as O(n).
f(n) = O(g(n)) is not same as O(g(n)) = f(n) (This is meaningless).

Example:

The function $3n + 2 = O(n)$ as $3n + 2 \leq 4n$ for all $n \geq 2$
$3n + 3 = O(n)$ as $3n + 3 \leq 4n$ for all $n \geq 3$

5 x + 10 is big-oh of $\mathbf{x^2}$, because $5x + 10 < 5x^2 + 10x^2 = 15x^2$ for $x > 1$.
Hence for $C = 15$ and $n_0 = 1$, $|5x + 10| \leq C|x^2|$.

Example:

$f(n) = 100 n^2$, $g(n) = n^4$, the following table and figure show that g(n) grows faster than f(n) when $n > 10$. We say f is big-Oh of g.

n	f(n)	g(n)
10	10,000	10,000
50	250,000	6,250,000
100	1,000,000	100,000,000
150	2,250,000	506,250,000

2.4.3 (Ω) Omega Notation

The Omega notation provides the lower bound for a specific function.

We can say that $f(n) = \Omega(g(n))$ iff there exist a +ve constant c and n_0 such that

$$f(n) \geq c * g(n) \text{ for all } n, n \geq n_0$$

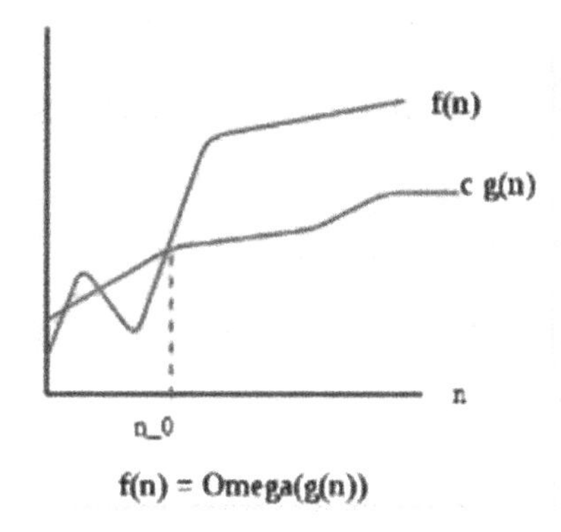

f(n) = Omega(g(n))

Example:

The function 3n + 2 = Ω (n) as 3n + 2 ≥ 3n for all n ≥ 1
3n + 3 = Ω (n) as 3n + 3 ≥ 4n for all n ≥ 1

2.4.4 (θ) Theta Notation

The Theta notation provides both the lower bound and upper bound for a specific function.

We can say that f(n) = θ (g(n)) iff there exist +ve constants c1 and c2 and n0 such that

$$C_1 * g(n) \le f(n) \le c_2 * g(n) \text{ for all } n, n \ge n_0$$

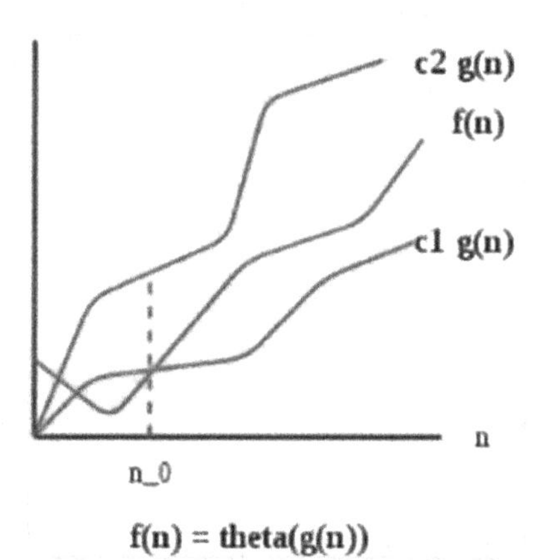

f(n) = theta(g(n))

Example:

The function $3n + 2 = \theta(n)$ as $3n + 2 \geq 3n$ and $\leq 4n$ for all $n \geq 2$, $c1 = 3$, $c2 = 4$.

2.4.5 o – Notation (Little–Oh Notation)

The function $f(n) = o(g(n))$ iff $\lim_{n\to\infty} \frac{f(n)}{g(n)} = 0$.

Example:

$3n + 2 = o(n^2)$ since $\lim_{n\to\infty} \frac{3n+2}{n^2} = 0$.
$3n + 2 = o(n \log n)$.

2.4.6 ω – Notation (Little-Omega Notation)

The function $f(n) = \omega(g(n))$ iff $\lim_{n\to\infty} \frac{g(n)}{f(n)} = 0$ OR $\lim_{n\to\infty} \frac{f(n)}{g(n)} = \infty$.

2.4.7 Comparison of Functions

2.4.7.1 Transitivity

$f(n) = \theta(g(n))$ and $g(n) = \theta(h(n)) \Rightarrow f(n) = \theta(h(n))$
$f(n) = O(g(n))$ and $g(n) = O(h(n)) \Rightarrow f(n) = O(h(n))$
$f(n) = \Omega(g(n))$ and $g(n) = \Omega(h(n)) \Rightarrow f(n) = \Omega(h(n))$
$f(n) = o(g(n))$ and $g(n) = o(h(n)) \Rightarrow f(n) = o(h(n))$
$f(n) = \omega(g(n))$ and $g(n) = \omega(h(n)) \Rightarrow f(n) = \omega(h(n))$

2.4.7.2 Reflexivity

$f(n) = \theta(f(n))$
$f(n) = O(f(n))$
$f(n) = \Omega(f(n))$

2.4.7.3 Symmetry

$f(n) = \theta(g(n))$ iff $g(n) = \theta(f(n))$

2.4.7.4 Transpose Symmetry

$f(n) = O(g(n))$ iff $g(n) = \Omega(f(n))$
$f(n) = o(g(n))$ iff $g(n) = \omega(f(n))$

2.4.8 Summary

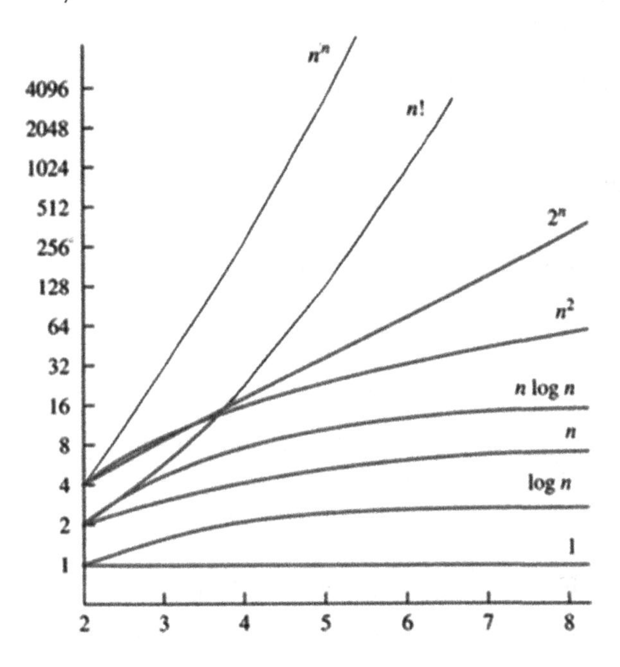

2.5 PROBLEMS RELATED TO NOTATIONS

Q Let f(n) and g(n) be asymptotically non negative functions then show that MAX(f(n),g(n)) = θ (f(n) + g(n))

Proof

Let $f(n) = An^2 + Bn + C$

$g(n) = Xn^3 + Yn^2 + Z$

$$\text{MAX}\left(f(n), g(n)\right) = \text{MAX}\left(An^2 + Bn + C,\ Xn^3 + Yn^2 + Z\right)$$

$$= Xn^3 + Yn^2 + Z = q(n^3) \tag{2.1}$$

$$\theta\left(f(n) + g(n)\right) = \theta\left(An^2 + Bn + C + Xn^3 + Yn^2 + Z\right)$$

$$= \theta\left(Xn^3 + (A+Y)n^2 + Bn + C + Z\right) \tag{2.2}$$

$$= \theta\left(n^3\right)$$

So from Equations 2.1 and 2.2

MAX(f(n), g(n)) = θ (f(n) + g(n)) **(Proved)**

Q Show that $2^{N+1} = O(2^N)$

Proof:

$2^{n+1} = 2 \cdot 2^n = O(2^n)$ where C = 2 **(Proved)**

Q Show that

$$\sum_{i=1}^{n} i2 \text{ is } O(n^3)$$

Proof:

$$\sum_{i=1}^{n} i2 = \frac{n(n+1)(2n+1)}{6} = \frac{(2n3+3n2+n)}{6} = O(n^3) \text{ (Proved)}$$

Q Show that if f(n) is O(g(n)) and d(n) is O(h(n)), then f(n) + d(n) is O(g(n) + h(n))

Proof:

As $f(n) = O(g(n))$ then $f(n) \le C.\ g(n)\ \forall\ C>0$

$d(n) = O(h(n))$ then $d(n) \le K.\ h(n)\ \forall\ K>0$

So $f(n) + d(n) \le C.\ g(n) + K.\ h(n)$

$\le A(g(n) + h(n))\ \forall\ A>0$

So $f(n) + d(n)$ is $O(g(n) + h(n))$ **(proved)**

Q Prove that lg(n!) = O(n lgn)

Proof:

$n! = n(n-1)(n-2) \ldots 1$

$\ge (n/2)^{n/2}$

Taking log on both sides

$\log n! \ge (n/2)^{n/2} = n/2(\lg n - \lg 2)$

$= n/2(\lg n - 1)$

$= n/2 \log n - n/2$

$\le n \log n$

$= O(n \lg n)$ **(proved)**

2.5.1 Stirling's Approximations

$$n! = \sqrt{2\Pi n}\,(n/e)^n\left(1+\theta\left(\frac{1}{n}\right)\right) \Rightarrow n! \cong (n/e)^n$$

Q Prove that lg(n!) = θ(n lgn)

Proof:

Lg n! = lg(n/e)^n = n lg(n/e)

= n lg n – n lg e

$$\lim_{n\to\infty}\frac{f(n)}{g(n)} = \frac{n\lg n - n\lg e}{\lim\limits_{n\to\infty} n\lg n} = 1 - \frac{\lg e}{\lg n} = 1$$

$$\therefore\ \lg n! \cong \theta(n\lg n)\ \ \textbf{(proved)}$$

Exercise Prove $\omega(g(n)) \cap O(g(n)) = \varnothing$, whence $\omega(g(n)) \subseteq \Omega(g(n)) - \Theta(g(n))$.

The following picture emerges:

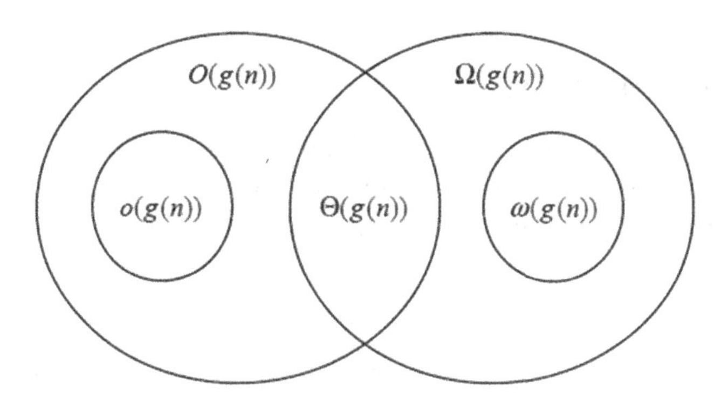

Theorem – 1

If f (n) = $a_m n^m + \ldots + a_1 n + a_0$, then f(n) = O($n^m$)

Proof:

f(n) = $a_m n^m + \ldots + a_1 n$ + a0

$$\leq \sum_{i=0}^{m} |a_i| n^i$$

$$\leq n^m \sum_{i=0}^{m} |a_i| n^{i-m}$$

$$\leq n^m \sum_{i=0}^{m} |a_i| \quad \text{for } n \geq 1$$

So f (n) = $O(n^m)$ **(proved)**

Theorem – 2

If $f(n) = a_m n^m + \ldots + a_1 n + a_0$ and $a_m > 0$ then $f(n) = \Omega(n^m)$

Proof:

$f(n) = a_m n^m + \ldots\ldots\ldots + a_1 n + a_0$

$= a_m n^m + (a_{m-1} n^{m-1} + \ldots\ldots\ldots + a_1 n + a_0)$

$\geq a_m n^m \; \forall \; a_m > 0$ and $n \geq 1$

Where $c = a_m$ and $n_0 > 0$

So $f(n) = \Omega(n^m)$

Proved

Example: Prove that $\sqrt{n+10} = \Theta(\sqrt{n})$.

Proof: According to the definition, we must find positive numbers c_1, c_2, n_0, such that the inequality $0 \leq c_1\sqrt{n} \leq \sqrt{n+10} \leq c_2\sqrt{n}$ holds for all $n \geq n_0$. Pick $c_1 = 1$, $c_2 = \sqrt{2}$, and $n_0 = 10$. Then if $n \geq n_0$, we have:

$$-10 \leq 0 \quad \text{and} \quad 10 \leq n$$

$$\therefore \; -10 \leq (1-1)n \quad \text{and} \quad 10 \leq (2-1)n$$

$$\therefore \; -10 \leq \left(1-c_1^2\right)n \quad \text{and} \quad 10 \leq \left(c_2^2-1\right)n$$

$$\therefore \; c_1^2 n \leq n+10 \quad \text{and} \quad n+10 \leq c_2^2 n,$$

$$\therefore \; c_1^2 n \leq n+10 \leq c_2^2 n,$$

$$\therefore \; c_1\sqrt{n} \leq \sqrt{n+10} \leq c_2\sqrt{n},$$

as required.

Lemma: $f(n) = o(g(n))$ if and only if $\lim_{n\to\infty} \frac{f(n)}{g(n)} = 0$.

Proof: Observe that $f(n) = o(g(n))$ if and only if $\forall c > 0, \exists n_0 > 0, \forall n \geq n_0 : 0 \leq \frac{f(n)}{g(n)} < c$, which is the very definition of the limit statement $\lim_{n\to\infty} \frac{f(n)}{g(n)} = 0$.

Example: $\lg(n) = o(n)$ since $\lim_{n\to\infty} \frac{\lg(n)}{n} = 0$. (Apply l'Hopital's rule.)

Lemma: If $\lim_{n\to\infty} \frac{f(n)}{g(n)} = L$, where $0 \leq L < \infty$, then $f(n) = O(g(n))$.

Proof: The definition of the above limit is $\forall \varepsilon > 0, \exists n_0 > 0, \forall n \geq n_0 : \left| \frac{f(n)}{g(n)} - L \right| < \varepsilon$. Thus, if we let $\varepsilon = 1$, there exists a positive number n_0 such that for all $n \geq n_0$:

$$\left| \frac{f(n)}{g(n)} - L \right| < 1$$

$$\therefore \quad -1 < \frac{f(n)}{g(n)} - L < 1$$

$$\therefore \quad \frac{f(n)}{g(n)} < L + 1$$

$$\therefore \quad f(n) < (L+1) \cdot g(n).$$

Now take $c = L + 1$ in the definition of O so that $f(n) = O(g(n))$ as claimed.

Lemma: If $\lim_{n\to\infty} \frac{f(n)}{g(n)} = L$, where $0 < L \leq \infty$, then $f(n) = \Omega(g(n))$.

Proof: The limit statement implies $\lim_{n\to\infty} \frac{g(n)}{f(n)} = L'$, $L' = 1/L$, and hence $0 \leq L' < \infty$. By the previous lemma $g(n) = O(f(n))$, and therefore $f(n) = \Omega(g(n))$.

Example: Compute the time complexity for the below recursive code

```
int fib(int n)
 {
   if((n==1) ||(n==2))return 1;
   return (fib(n-1)+fib(n-2));
 }
```

Converting the above code into a Recurrence Relation:

```
T(0) = 0 ......base case
T(n) = 2T(n-1) + 1
```

Solution

```
T(n) = 2T(n-1) + 1
     = 2(2T(n-2) + 1) + 1
     = 4(2T(n-3) + 1) + 1 + 2
     :
     :
     = 2^k T(n-k) + ∑ 2^i
     = 2^k T(n-k) + 2^(k-1)
     = 2^k T(n-k) + (2^k) - 1          using 1-r^n / 1-r
```

```
Now,

n-k = 0
n=k

Finally,

T(n) = 2^n T(n-n) + 2^n - 1
     = 2^n T(0) + 2^n - 1
     = 0 + 2^n - 1
     = 2^n - 1
Order:

O(2^n)  .....Time Complexity
```

2.6 RECURRENCES

2.6.1 Recurrence Relations

While analyzing the run time of recursive algorithms, we are often required to consider functions T(n), which are defined by recurrence relations of a certain form. A typical example would be

$$T(n) = \begin{cases} c & n = 1 \\ T(\lfloor n/2 \rfloor) + T(\lceil n/2 \rceil) + dn & n > 1 \end{cases}$$

where c and d are fixed constants. The specific values of these constants are important for determining the explicit solution to the recurrence. Often, however, we are only concerned with finding an asymptotic (upper, lower, or tight) bound on the solution. We call such a bound an *asymptotic solution* to the recurrence. In the above example, the particular constants c and d have no effect on the asymptotic solution. We may therefore write our recurrence as

$$T(n) = \begin{cases} \Theta(1) & n = 1 \\ T(\lfloor n/2 \rfloor) + T(\lceil n/2 \rceil) + \Theta(n) & n > 1 \end{cases}$$

Subsequently, we'll show that a tight asymptotic solution to the above recurrence is T(n) = Q(nlogn).

We will study the following methods for solving recurrences.

1. **Substitution Method.** This method consists of guessing an asymptotic (upper or lower) bound on the solution and trying to prove it by induction.
2. **Recursion Tree Method – Iteration Method.** These are two closely related methods for expressing T(n) as a summation, which can then be analyzed. A recursion tree is a graphical depiction of the entire set of recursive invocations of the function T. The goal of drawing a recursion tree is to obtain a guess, which can then be verified by a more rigorous substitution method. Iteration consists of repeatedly substituting the recurrence into itself to obtain an iterated (i.e., summation) expression.
3. **Master Method.** This is a cookbook method for determining asymptotic solutions to recurrences of a specific form.

2.6.2 Substitution Method

The substitution method for solving recurrences consists of two steps:

- Guess the form of the solution.
- Use mathematical induction to find constants in the form and show that the solution works.

The inductive hypothesis is applied to smaller values, similar to recursive calls bring us closer to the base case.

The substitution method is powerful to establish lower or upper bounds on a recurrence.

1. Prove by Substitution method that $T(n) \le 2C\lfloor n/2 \rfloor \log(\lfloor n/2 \rfloor) + n$

Show that $T(n) \le c.n \log n$ for large enough c and n.

Solution

Let us assume that it is true for (n/2), then:

$T(n) \le 2c\lfloor n/2 \rfloor \log(\lfloor n/2 \rfloor) + n$

$\le c.n. \log(|n/2|) + n$

$= c.n(\log n - \log 2) + n$

$= c.n \log n - cn + n$ [$\log 2 = 1$]

$\le c.n\log n \ \forall\ n > 1$ [$T(1) = 0$]

Now

$T(2) \le C2\log 2$

$T(3) \le C3\log 3$

As n cannot be 1 so this relation holds for $c \ge 2$.

Thus $T(n) \le c.n. \log n$ and $T(n) = O(n.\log n)$ is true.

2. Show by substitution method that $T(n) = T(\lfloor n/2 \rfloor) + 1$ is $O(\log n)$

Proof: Let us guess that solution is $O(\log n)$.

So we have to prove that $T(n) \le C \log n$

By substituting this we get

$T(n) \le c \log(n/2) + 1$

$= c \log n - c \log 2 + 1$

$= c \log n - c + 1$

$\le c \log n$ for $c >= 1$

So $T(n) = O(\log n)$ **Proved**

3. Show that solution to $T(n) = 2T((\lfloor n/2 \rfloor/2) + 17) + n$ is $O(n \lg n)$

Solution: Let us guess that solution is $O(n \lg n)$ then we have to prove $T(n) \le cn\lg n$

By substituting

$$T(n) \le 2\left(\frac{cn}{2}\lg\left(\frac{n}{2}\right)+17\right)+n$$

$$= cn\lg^{\frac{n}{2}} + 34 + n$$

$$= cn\lg n - cn\lg 2 + 34 + n$$

$$= cn\lg n - cn + 34 + n$$

$$= cn\lg n - (c-1)n + 34$$

$$= cn\lg n - bn + 34$$

$$\le cn\lg n$$

If $c \ge 1$, $T(n) = O(n \lg n)$.

4. Show that $T(n) = 2T(\lfloor n/2 \rfloor) + n$ is $\Omega(n \lg n)$

Solution: Let T(n) is $\boldsymbol{\Omega(n \lg n)}$, then we have to prove $T(n) \ge cn \lg n$.

$$T(n) \ge 2(c\frac{n}{2}\frac{\lg n}{2}) + n$$

$$\ge cn\frac{\lg n}{2} + n$$

$$= cn\lg n - cn\lg 2 + n$$

$$= cn\lg n - cn + n$$

$$\ge cn\lg n \text{ for } c \ge 1$$

$\therefore$ $T(n) = \Omega(n \lg n)$ for $c \le 1$

5. Consider the recurrence

$$T(n) = 2T\left(\left\lfloor\sqrt{n}\right\rfloor\right) + \log_2(n)$$

Rename $m = \log_2(n)$. We have:

$$T(2^m) = 2T(2^{m/2}) + m.$$

Define $S(m) = T(2^m)$. We get:

$$S(m) = 2S(m/2) + m.$$

Hence, the solution is $O(m \log_2(m))$, or with substitution

$$O(\log_2(n) \cdot \log_2(\log_2(n))).$$

2.6.3 Recursion Tree

Recursion tree is another way of solving the recurrences. Drawing a recursion tree can be thought of as a good guess for a substitution method. In the recursion tree, each node represents the cost of a single subproblem somewhere in the set of recursive function invocations. We sum up the costs within each level of the tree to obtain a set of per level costs, and then we sum up all per level costs to determine the total cost of all the levels of the recursion. This method is helpful for the divide-and-conquer algorithms.

The *recursion tree method* can be used to generate a good guess for an asymptotic bound on a recurrence. This guess can then be verified by the substitution method. Since our guess will be verified, we can take some liberties in our calculations, such as dropping floors and ceilings or restricting the values of n.

Let us consider the example

$$T(n)=\begin{cases}\Theta(1) & 1\leq n<3\\ 2T(\lfloor n/3 \rfloor)+\Theta(n) & n\geq 3\end{cases}$$

We simplify the recurrence by dropping the floor and replacing Q(n) with n to get T(n) = 2T(n/3) + n.

Each node in a recursion tree represents one term in the calculation of T(n) obtained by recursively substituting the expression for T into itself. We construct a sequence of such trees of increasing depths.

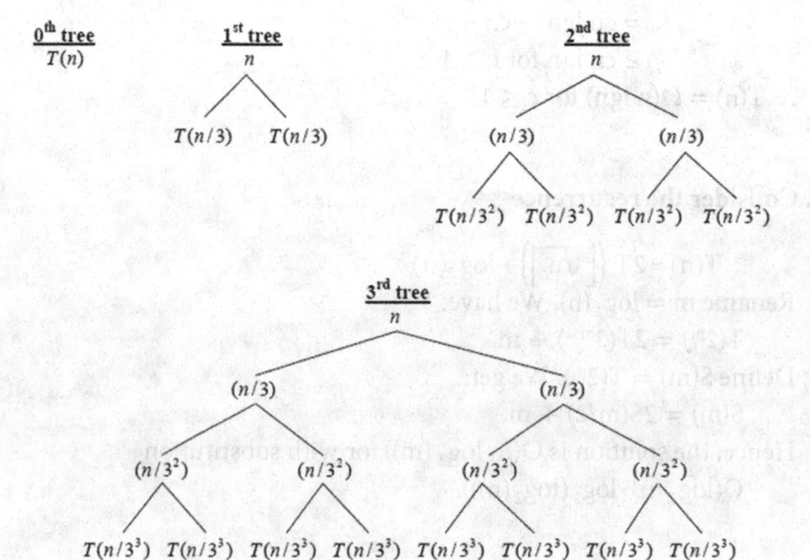

By summing the nodes in any one of these trees, we obtain an expression for T(n).

After k iterations of this process, we reach a tree in which all bottom level nodes are $T(n/3^k)$.

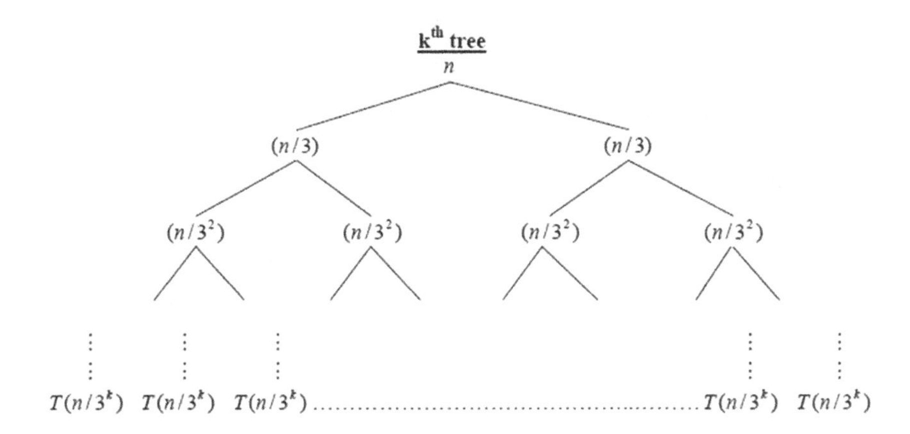

Note that there are 2^i nodes at depth I, each of which has the value $n/3^i$ (for $0 \le i \le k-1$). The sequence of trees terminates when all bottom level nodes are T(1), i.e., when $n/3^i = 1$, which implies $k = \log_3(n)$. The number of nodes at this bottom level is therefore $2^k = 2^{\log 3\,(n)} = n^{\log 3\,(2)}$. Summing all the nodes in the

$$T(n) = \sum_{i=0}^{k-1} 2^i \cdot (n/3) + n^{\log_3(2)} \cdot T(1)$$

$$= n\left(\sum_{i=0}^{k-1} (2/3)^i\right) + n^{\log_3(2)} \cdot T(1)$$

$$= n\left(\frac{1-(2/3)^k}{1-(2/3)}\right) + n^{\log_3(2)} \cdot T(1)$$

$$= 3n\left(1-(2/3)^k\right) + \Theta\left(n^{\log_3(2)}\right)$$

If we seek an asymptotic upper bound, we may drop the negative term to obtain $T(n) \le 3n + \Theta\left(n^{\log_3(2)}\right)$. Since $\log_3(2) < 1$, the first term dominates, and so we guess: T(n) = O(n).

Given T(n) = 2T(n/2) + n, the recursion tree for the relation will be

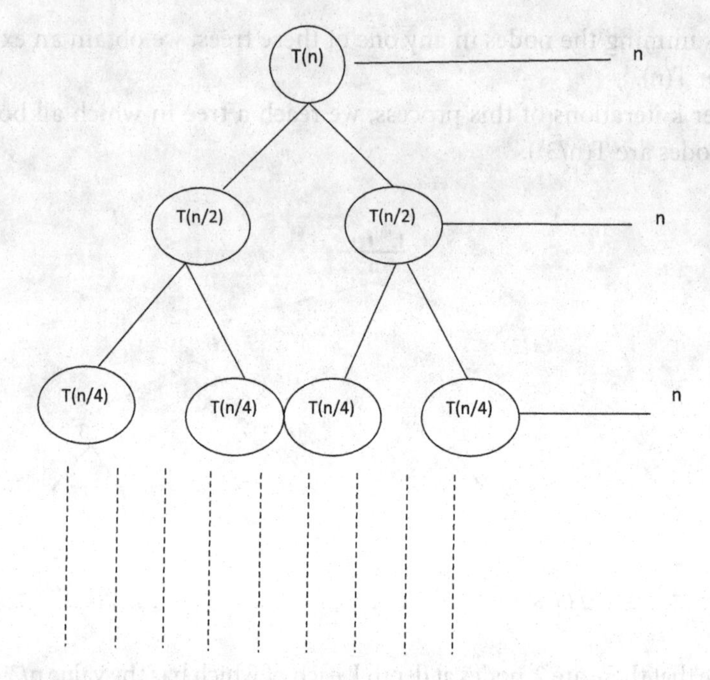

To compute the cost of the above tree, we have to find the cost of each level and then add them.

Therefore, the cost of level 0 is n

The cost of level 1 is n

The cost of level 2 is n and so on…

Now compute the number of levels in the tree.

Size of subproblem at level 0 is $n/2^0$

Size of subproblem at level 1 is $n/2^1$

Size of subproblem at level 2 is $n/2^2$

……………………………………………………………………………

……………………………………………………………………………

Size of subproblem at level i is $n/2^i$

Let at level m size of subproblem will be 1

Then $n/2^m = 1$

Therefore, $2^m = n$

Taking logarithm both sides

$M\log_2 = \log n$

$M = \log_2 n$

Therefore, the total number of levels in the recursion tree is $\log_2 n + 1$.

Therefore, the total cost will be

(Total number of nodes) * $\log_2 n$

Now compute total number of nodes

At level-0 nodes = 2^0 = 1 node.
At level-1 nodes = 2^1 = 2 nodes.
At level-2 nodes = 2^2 = 4 nodes.

Since we have $\log_2 n$ number of levels, i.e., we have $2^{\log 2n}$ nodes, i.e., n nodes

Therefore, the cost of last level = n * T(1) = O(n)

Now add the cost of each levels.

$$T(n) = n + n + n + \dots + O(n)$$
$$= n * \log_2 n + O(n)$$
$$= O(n \log_2 n)$$

Example: Draw the recursion tree for $T(n) = 2T(n/2) + n^2$.

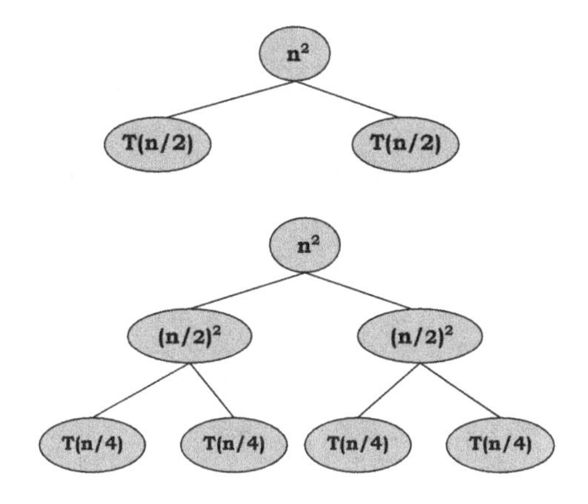

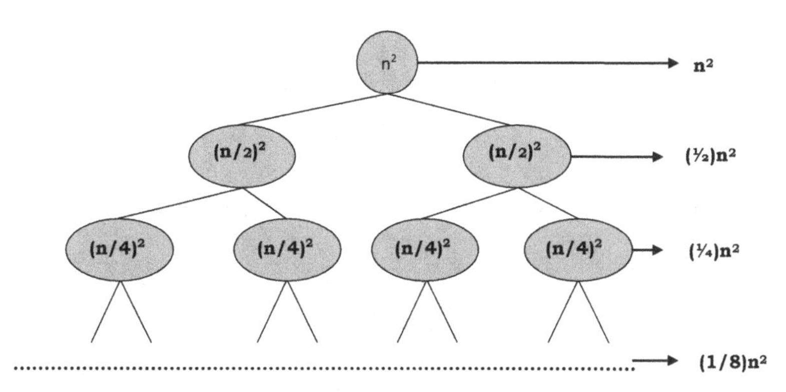

Summing the values at each level, we get

$$n^2 + (½)n^2 + (¼)n^2 + \ldots = n2\left(a + \frac{1}{2} + \frac{1}{4} + \ldots\right)$$

Hence, the solution is $\Theta\left(n^2\right)$.

Example

Consider the recurrence $T(n) = 3T\left(\left\lfloor \frac{n}{4} \right\rfloor\right) + cn^2$.

Simplification: We assume that n is a power of 4.

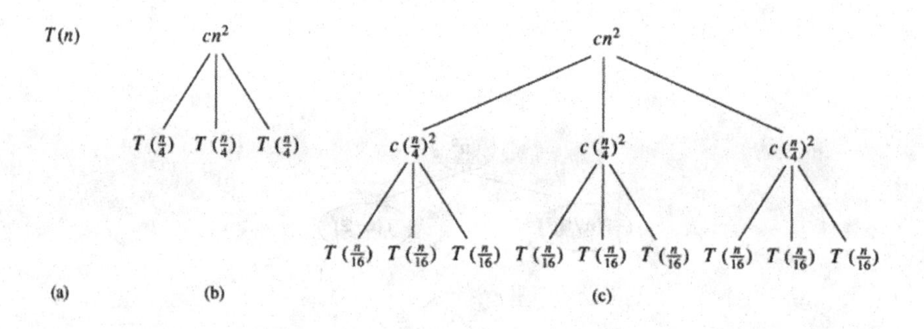

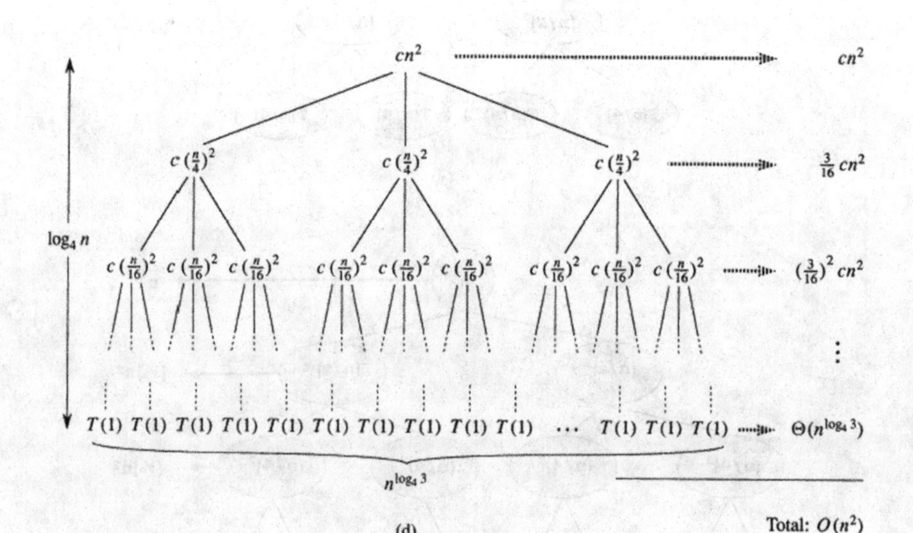

Example

$$T(n) = T\left(\frac{n}{3}\right) + T\left(\frac{2n}{3}\right) + cn$$

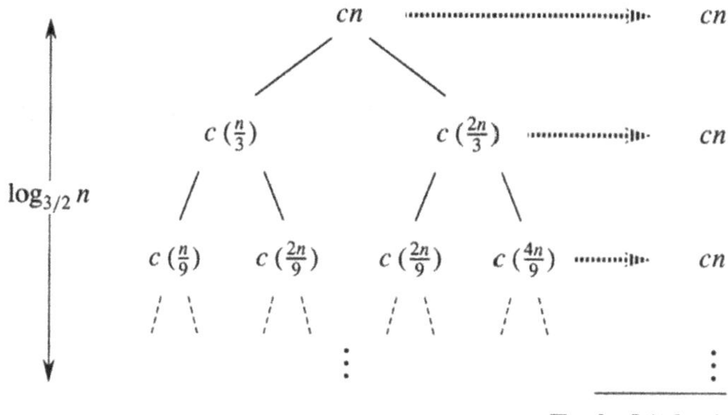

2.6.4 Master Method

The simplest form of the Master method is

$$T(n) = a\,T(n/b) + \theta(n^c)$$

To solve these type of recurrence relations, we have to go through three different cases as

Case 1	$c < \log_b^a$	$T(n) = \theta\,(n^{\log_b a})$
Case 2	$c = \log_b^a$	$T(n) = \theta\,(n^c{}_{\log n}) = \theta\,(n^{\log_b a} \cdot \log n)$
Case 3	$c > \log_b^a$	$T(n) = \theta\,(n^c)$

The general form of the Master method is

Let a ≥ 1 and b >1 be constants, let f(n) be a function and Let T(n) be defined on the non-negative integers by recurrence

$$T(n) = a\,T\left(\frac{n}{b}\right) + f(n)$$

where

$\left(\frac{n}{b}\right)$ can be ceil or floor.

Then T(n) can be asymptotically bounded as

1. If $f(n) = O(n^{\log_b a-\varepsilon})$ for some constant $\varepsilon > 0$, then $T(n) = \theta(n^{\log_b a})$.
2. If $f(n) = \theta(n^{\log_b a})$, then $T(n) = \theta(n^{\log_b a} \cdot \log n)$.
3. If $f(n) = \Omega(n^{\log_b a+\varepsilon})$ for some constant $\varepsilon>0$ and if $af\left(\frac{n}{b}\right) \leq c\,f(n)$ for some constant $c < 1$ and for all sufficiently large n, then $T(n) = \theta(f(n))$

Example:

$T(n) = 4\,T(n/2) + \theta(n^3)$
Here a = 4, b = 2, c = 3
$\log_b a = \log_2 4 = 2$
Since $c > \log_b a$ So $T(n) = \theta(n^c)$ i.e./$T(n) = \theta(n^3)$

Example:

$T(n) = 4\,T(n/2) + \theta(1)$
Here a = 4, b = 2, c = 0
$\log_b a = \log_2 4 = 2$
Since $c < \log_b a$, $T(n) = \theta(n\log_b a)$, i.e./ $T(n) = \theta(n^2)$,

Example 1: Solve $T(n) = 4\,T\left(\frac{n}{2}\right) + n$
According to the Master method
A=4 b = 2 f(n) = n thus
$n\log_b a = n\log_2 4 = n^2$
Thus, we are in case 1
For f(n) is $O(n^{2-\varepsilon})$ for $\varepsilon = 1$
So $T(n) = \theta(n^2)$.

Example 2: Solve $T(n) = 2T\left(\frac{n}{2}\right) + n$
Solution: a = 2, b = 2, f(n) = n

$n\log_b a = n\log_2 2 = n$
Since $f(n) = n\log_b a$ so from case 2, we have
$T(n) = \theta(n \log n)$

Example 3: Solve $T(n) = 3T\left(\frac{n}{4}\right) + n \log n$

Solution: a= 3 , b = 4, f(n) = n logn

$$n^{\log_b a} = o(\mathbf{n^{0.793}})$$

Since here $f(n) = \Omega(n^{\log_b 3+\varepsilon})$ where $\varepsilon = 0.2$

Here case 3 holds, now we have to check the regularity condition holds for f(n) for sufficiently large value of n.

$(n/b) = 3\ (n/4)\log(n/4) \leq (3/4)\ n \log n = c.f(n)$ for $c = 3/4 < 1$

So $T(n) = \theta(n \log n)$.

Example 4: Solve $T(n) = T\left(\frac{2n}{3}\right) + 1$

Here $a = 1$, $b = \frac{3}{2}$ $f(n) = 1$

$n^{\log_b a} = n^{\log_{(3/2)} 1} = n^0 = 1$

So from case 2 $T(n) = \theta(\log n)$.

Example 5: Solve $T(n) = 9T\left(\frac{n}{3}\right) + n$

Here $a = 9$, $b = 3$ $f(n) = n$

$n^{\log_b a} = n^{\log_{(3)} 9} = \theta(n^2)$

$f(n) = O(n^{\log_{(3)} 9} - 1) = O(n)$

So from case 1 $T(n) = \theta(n^2)$

Example 6: Solve $T(n) = 2T\left(\frac{n}{2}\right) + n^3$

Here $a = 2$, $b = 2$ $f(n) = n^3$

$n^{\log_b a} = n^{\log_{(2)} 2} = n$

$f(n) = \Omega(n^{\log_{(2)} 2+1})$

Now $2f(n^3 / 2^3) \leq c\ f(n^3)$ is true for $c = 1/4$

Hence, the regularity condition holds so $T(n) = \theta(n^3)$.

Example 7: Solve $T(n) = T\left(\frac{9n}{10}\right) + n$

Here $a = 1$, $b = \frac{10}{9}$ $f(n) = n$

$n^{\log_b a} = n^0 = 1$
$f(n) = \theta(n^{0+1})$
So case 3 holds
$af\left(\frac{n}{b}\right) \leq cn$
i.e./$f\left(\frac{9n}{10}\right) \leq c(n)$ for c = 9/10
Hence, the regularity condition holds so $T(n) = \theta(n)$.

Example 8: Solve $T(n) = 16T\left(\frac{n}{4}\right) + n^3$

Here a = 16, b =4 f(n) =n^3
$n\log_b{}^a = n\log_4{}^{16} = n^2$
$f(n) = \theta(n^{2+1}) = \Omega(n^3)$
So case-III holds
$af\left(\frac{n}{b}\right) \leq cn$ i.e./$16f(n^3/4^3) \leq c\, f(n^3)$ where c = ¼
Hence, the regularity condition holds so $T(n) = \theta(n^3)$.

Example 9: $T(n) = T\left(\sqrt{n}\right) + 1$

Ans: Let m = log n so $n = 2^m$
$\Rightarrow T(2^m) = T(2^{m/2}) + 1$
$S(m) = S\left(\frac{m}{2}\right) + 1$ [$\therefore T(2^m) = S(m)$]
Here a=1, b = 2, f(m) =1
$m^{\log_b a} = m^{\log_2 1} = 1 = \theta(1)$
So by case-II $S(m) = \theta(m^{\log_b a \cdot \log m})$
$S(m) = \theta(1 \cdot \log m)$
$S(m) = \theta(\log \cdot \log n)$
$T(n) = T(2^m) = S(m) = \theta(\log \cdot \log n)$

Example 10: $T(n) = 2T\left(\frac{n}{4}\right) + \sqrt{n}$

Ans: a = 2 b = 4 $f(n) = n^{(1/2)}$
$n^{\log_b a} = n^{\log_4 2} = n^{(1/2)} = O(n^{(1/2)})$
So from case-II $f(n) = \theta(n^{\log_b a \cdot \log n})$
$= \theta(n^{(1/2).\log m})$
$\therefore T(n) = \theta\ (\sqrt{n} \log n)$

2.7 QUESTIONS

2.7.1 Short questions

1. What are algorithm testing and its type?
2. What is time complexity? What is the notation to express it?
3. What shows the order of growth of a notation?
4. How many types of asymptotic notations are there? Write all the notations.
5. How many types of methods are used in recurrence relation?
6. Write the names of all methods.
7. What is an algorithm?
8. What do you mean by algorithm validation?
9. Ordering asymptotic growth rates of the following n^2, 2^{logn}, (logn)!, n^3, nlogn.
10. If $f(n) = 5n^2 + 6n + 4$, then prove that f(n) is $O(n^2)$.
11. Why an asymptotic notation is used?

2.7.2 Long Questions

1. Draw the recursion tree for $T(n) = 2T(n/2)+n^2$.
2. Solve $T(n) = 4T\ n/(2)+n$ through the Master method.
3. Prove by substitution method that $T(n) < 2c[n/2] \log([n/2]) + n$. Also show that $T(n) < c.\ n\log n$ for large values of c and n.
4. Draw the recurrence tree for $T(n) = 8T(n/2) + n_2(T(1) = 1)$.
5. Consider the recurrence $T(n) = 3T([n/4]) + cn^2$.
6. Define all the notations with figure.
7. Write transitive property for all notations.
8. Prove that $MAX(f(n), g(n)) = 0(f(n)+g(n))$.

9. Show that if f(n) is O(g(n)) and d(n) is O(h(n)), then f(n)+d(n) is O(g(n)+h(n)).

10. Solve $\mathbf{T(n) = T\left(\sqrt{n}\right) + 1}$.

CHAPTER 3

Divide-and-Conquer Techniques

3.1 DIVIDE-AND-CONQUER APPROACH

Divide and conquer is a top-down technique for designing algorithms that consist of dividing the problem into smaller subproblems, hoping that the solutions to the subproblems are easier to find and then composing the partial solutions into the solution of the original problem.

Little more formally, the divide-and-conquer paradigm consists of the following major phases:

- Breaking the problem into several subproblems that are similar to the original problem but smaller in size,
- Solve the subproblem recursively (successively and independently), and then
- Combine these solutions with subproblems to create a solution to the original problem.

3.2 BINARY SEARCH

Let A_i, $1 \leq i \leq n$, be a list of elements that are sorted in increasing order. Consider the problem of determining whether a given element x is present in the list.

DOI: 10.1201/9781003093886-3

If x is present, we are to determine a value J such that Aj=x.

If x is not in the list, then J is to be set to zero.

Let P=(n, A_i ... A_l, x) denote an arbitrary instance of this search problem.

n is the number of elements in the list as A_i ... A_l and x is the element to search.

For binary search, the concept of divide and conquer can be used.

Let small(p) be true if n=1, so in this case, S(P) will take the value i if x=Ai.

Otherwise, it will take the value to zero.

In binary search, mainly three possibilities are used:

- If x=Aq, then our search problem is successfully completed.
- If x<Aq, then here x has to be searched for only the sublist A_i, $A_{(i+1)}$ ... $A_{(q-1)}$.
- ∴P reduces to (q−i, Ai, ... Aq−1, x).
- If x>Aq, then here x has to be searched for only the sublist $A_{(q+1)}$, ... $A_{(l)}$.
- ∴P reduces to (l−q, $A_{(q+1)}$, ... A_l, x).

  1. The binary search method can be implemented in two ways. Recursive
  2. Iterative

Recursive

```
 1. ALGORITHM BinSearch(A,i,l,x) [A is the array, i is
    the Lower bound]
 2. {
 3. If(l=i) then    [if small(p)]
 4. {
 5. If(x = A[i]) then return i;
 6. else
 7. return 0;
 8. }
 9. else
10. { [reduce P into a smaller problems]
11. mid ← ⌊(i+l)/2⌋
```

```
12. if (x = A[mid]) then return mid;
13. else
14. if(x< A[mid]) then
15. return BinSearch(A,i,mid-1,x);
16. else
17. return BinSearch(A,mid+1,l,x);
18. }
19. }
```

Iterative

```
 1. ALGORITHM BinSearch(A,n,x)
 2. {
 3. low ← 1
 4. high ← n
 5. while(low ≤ high ) do
 6. {
 7. mid ← ⌊(low+high)/2⌋
 8. If (x < A[mid]) then high := mid-1
 9. else
10. if (x > A[mid]) then low := mid+1
11. else
12. return mid
13. }
14. return 0
15. }
```

3.2.1 Analysis of Binary Search

For analysis, first we have to compute the number of comparisons expected for a successful binary search.

Therefore, consider i such that

$$2^i \geq (N+1)$$

Thus, $2^{(i-1)} - 1$ is the maximum number of comparisons that are left with the first comparison.

Similarly, $2^{(i-2)} - 1$ is the maximum number of comparisons that are left with the second comparison.

In general, we can say that $2^{(i-k)} - 1$ is the maximum number of comparisons that are left after K comparisons.

If no elements are left, then the desired data item is obtained by ith comparisons.

3.2.1.1 *Best-Case complexity*

If the searched value is found at the middle of the list, then the comparison required $T(n)=O(1)$.

3.2.1.2 *Worst-Case complexity*

Let K be the smallest integer such that $n<=2^k$ and c is one constant time required for one comparison, so

$T(n)=T(n/2)+c$

$t(2^k)=T(2^K/2)+c \Rightarrow T(2^k)=T(2^{k-1})+c$

By the method of induction, we have

$T(2^k)=T(2^{k-1})+c$

$T(2^{k-1})=T(2^{k-2})+c$

……………………

……………………

……………………

……………………

$T(2^1)=T(2^{k-1(k-1)})+c$

$T(2^k)=T(1)+kc$

$T(n)<=kc$ T(1) as constant

$T(n)<=c^* \log_2 n$ ($n=2^k \Rightarrow k=\log_2 n$)

$T(n)=O(\log n)$

Finally we can say that for the best case O(1)

For the average case O(logN)

For the worst case O(logN)

3.3 MERGE SORT

In general, the merge sort works as

1. If the list is of the length 0 or 1, then it is already sorted, otherwise.
2. Divide the unsorted list into two sublists of about half the size.
3. Sort each sublist recursively by re-applying merge sort.
4. Merge the two sublists back into one sorted list.

Merge sort incorporates two main ideas to improve its runtime.

a. A small list will take fewer steps to sort than a large list.

b. Fewer steps are required to construct a sorted list from two sorted lists than two unsorted lists.

Algorithm

```
 1. ALGORITHM Merge Sort(l,h)
 2. {
 3. If (l < h) then
 4. {
 5. m ← (l+h)/2
 6. Merge Sort (l,m)
 7. Merge Sort(m+1,h)
 8. Merge(l,m,h)
```

```
 1. ALGORITHM Merge(l,m,h)
 2. {
 3. k ← l, I ← l, j ← m+1
 4. while ((k≤m) and (j≤h)) do
 5. {
 6. If(a[k] ≤ a[j]) then
 7. {
 8. b[i] ← a[k]
 9. k ← k+1
10. }
11. else
12. {
13. b[i] ← a[j]
14. j ← j+1
15. }
16. i ← i+1
17. }
18. If(k > m) then
19. for x = j to h do
20. {
21. b[i] ← a[x]
22. i ← i+1
23. }
24. else
25. for x = k to m do
26. {
27. b[i] ← a[x]
```

```
28. i ← i+1
29. }
30. for x = 1 to h do
31. a[x] ← b[x]
32. }
```

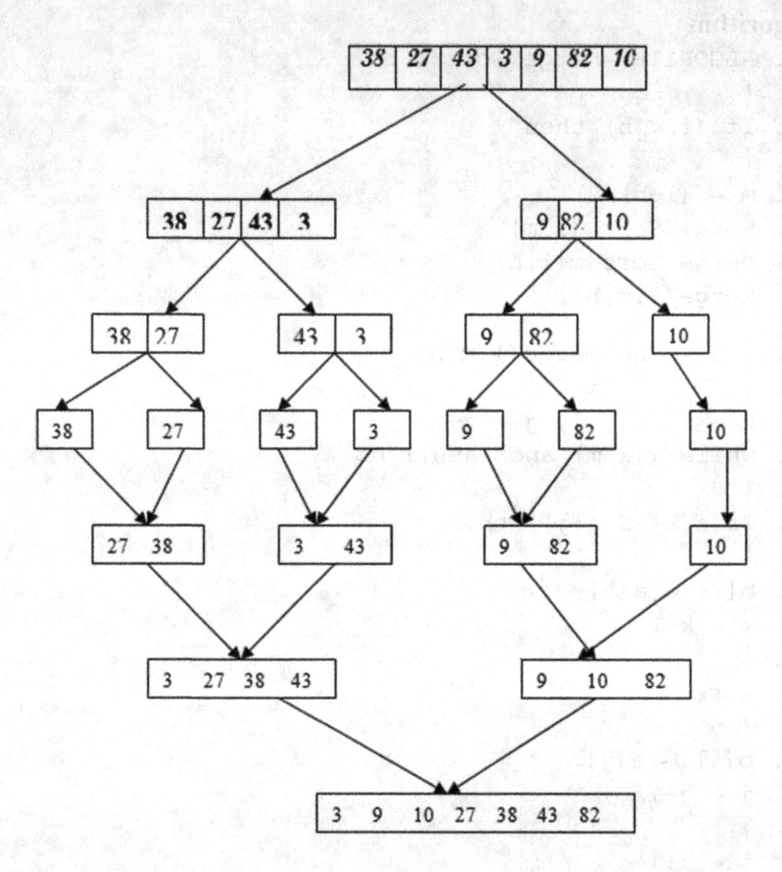

3.3.1 Analysis of Merge Sort

Let us analyze the running time of the entire merge sort algorithm assuming it is given an input sequence of n number of elements.

Let us consider n is a power of 2, now each divide step then yields two subsequences of size exactly n/2.

Let us set up the recurrence for T(n), the worst-case running time of merge sort of n numbers.

Merge sort for one element will take a constant time, but when we have n>1 elements, then we break the running time as

- **Divide**: The divide step just computes the middle of the subarray which take the constant time i.e./$T(n)=\theta(1)$.
- **Conquer**: We recursively solve two subproblems, each of size n/2 which will take $2T(n/2)$ to the running time.
- **Combine**: We know that the merge procedure on n elements subarray takes the time $\theta(n)$.

Therefore, for the worst case of merge sort, the T(n) can be represented as

$$T(n)=\theta(1) \text{ if } n=1$$

$$T(n)=2T(n/2)+n \text{ if } n>1$$

So from the Masters Theorem, the time $T(n)=\theta(n \log n)$.

3.4 QUICK SORT

The quick sort algorithm was invented by C.A.R Hoare in 1960 but formally introduced in 1962.

It is based on the principle of the divide-and-conquer method.

3.4.1 Good Points of Quick Sort

- It is in place since it uses only a small auxiliary stack.
- It requires only n log(n) time to sort n items.
- It has an extremely short inner loop.

3.4.2 Bad Points of Quick Sort

- It is recursive, especially if recursion is not available, the implementation is extremely complicated.
- It requires quadratic, i.e., n^2 time in the worst case.
- It is fragile, i.e., a simple mistake in the implementation can go unnoticed and cause it to perform badly.

By using the divide-and-conquer method

- **DIVIDE**: Partition the array A[P … R] into two subarrays A[P … q−1] and A[q+1 … R] such that each element of A[P … q−1] is less than or equal to A[q], which is, in turn, less than or equal to each element of A[q+1 ... R].

- **CONQUER**: Sort the two subarrays A[P ... q−1] and A[q+1 ... R] by recursive calls to Quick Sort.
- **COMBINE**: Since the subarrays are sorted in place, no work is needed to combine them. The entire array A[P ... R] is now sorted.

Example:

Sort the elements 5 14 2 9 21 34 17 19 1 44 by using quick sort.

Solution:

First, choose the first element as pivot and find the first element larger than pivot from the left-hand side and find the smallest number equal to pivot from right to left.

After finding that swap, the larger element with the smallest element and continue the above-said process until the smallest element was found after the largest element. Once the smallest element was on the left-hand side of the largest, interchange the smallest element with the pivot so that again the array will be subdivided into two subarrays and implement the process on both the subarrays.

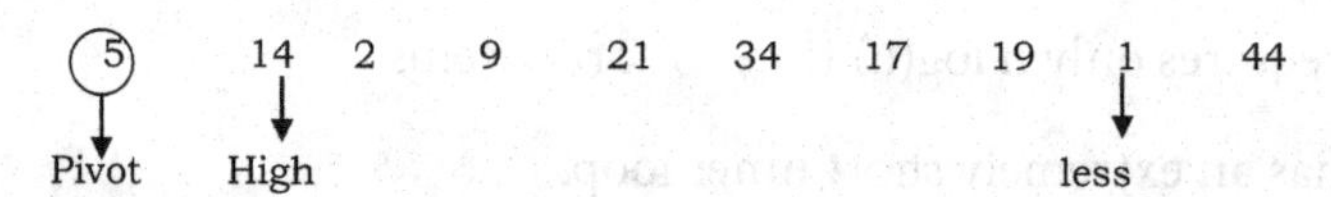

Swap 14 and 1

5 1 2 9 21 34 17 19 14 44

Less High

Since less < High so interchange the Pivot with less

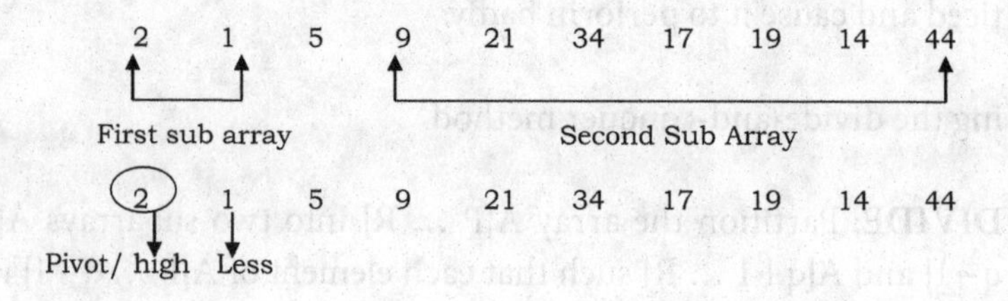

Since Pivot and high, both are the same so interchange the value of less with high

1 2 5 9 21 34 17 19 14 44

Since left subarray elements are sorted, now implement the sorting logic for the right subarray.

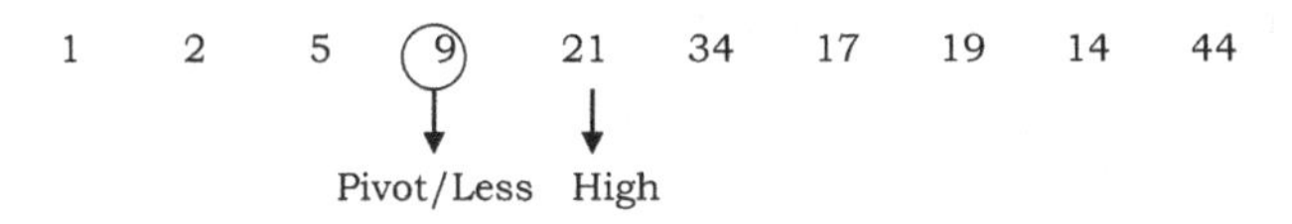

Since Pivot and less, both are the same so nothing will change, i.e., the left subarray having no elements, but the right subarray will have the elements from 21 to 44.

1 2 5 9 21 34 17 19 14 44

Pivot (21), High (34), Less (14)

1 2 5 9 21 14 17 19 34 44

Less (19), High (34)

1 2 5 9 19 14 17 21 34 44

Pivot/high (19), Less (17)

1 2 5 9 17 14 19 21 34 44

Pivot/high (17), Less (14)

1 2 5 9 14 17 19 21 34 44

Algorithm Quicksort(A,P,R)

```
1. If P<R
2. Then Q ← PARTITION(A,P,R)
3. QUICKSORT(A,P,Q)
4. QUICKSORT(A,Q+!,R)
```

Algorithm PARTITION(A,P,R)

```
1. x ← A[P]
2. low ← P
3. up ← R
```

```
 4. while(low < up)
 5. {
 6. while(A[low] < = x)
 7. low ← low+1
 8. while(A[up] > x)
 9. up ← up-1
10. if(low<up) then
11. exchange(A[low] ↔ A[up])
12. else
13. return(up)
14. }
```

3.4.3 Performance of Quick Sort

The running time of quick sort is completely dependent on the partition procedure.

3.4.3.1 Worst Case

It occurs when partitioning is such that at every stage, one subarray contains $n-1$ elements and the other contains one element.

Since partitioning for n number of elements time costs $\theta(n)$ and single element will take $\theta(1)$ so its recurrence relation can be represented as

$$T(n)=T(n-1)+\theta(n)$$

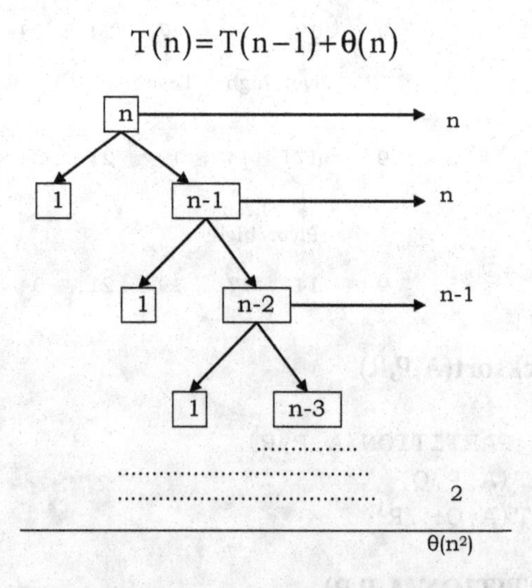

The Partition (A, P, R) call always return Q so successive calls to Partition will split arrays of length n, $n-1$, $n-2$, … 2 and running time proportional to

$$n+(n-1)+(n-2)+\ldots+2=\frac{(n+2)(n+1)}{2}=\theta(n^2)$$

3.4.3.2 Best Case

If partitioning is done in middle, it will be faster so

$$T(n)=2T\left(\frac{n}{2}\right)+\theta(n)$$

$$\therefore\ T(n)=\theta(n\log n)$$

3.5 HEAP SORT

- **Definition of Heap**:

  A max heap is a complete binary tree with the property that the value at each node is at least as large as (as small as) the values at its children (if they exist).

  This property is known as Heap Property.

  In general, max heap is usually used for heap sort and min heap is used in Priority Queue.

Theorem: A heap T storing n elements has height log (n + 1)

Proof:

Since T is complete, the number of internal nodes of T is at least

$1+2+4+\ldots+2^{h-2}+1=2^{h-1}$

This lower bound is achieved when there is only one internal node on level h - 1.

We observe that the number of internal nodes of T is at most

$1+2+\ldots+2^{h-1}=2^{h-1}$

This upper bound is achieved when all the 2^{h-1} nodes on level $h-1$ are internal.

Since the number of internal nodes is equal to the number n elements

$2^{h-1} <= n$ and $n <= 2^h - 1$

Thus, by taking logarithms of both sides of these two inequalities

$$h\le\log(n+1)\text{ and }\log(n+1)\le h$$

$$\Rightarrow h=\lceil\log(n+1)\rceil$$

Algorithm Max_Heap(A,i)

```
 1. {
 2. l ← left_child(index)
 3. r ← right_child(index)
 4. if l ≤ heapsize[A] and A[l]> A[i]
 5. then largest ← l
 6. else
 7. largest ← i
 8. if r ≤ heapsize[A] and A[r] > A[largest]
 9. then largest ← r
10. if largest ≠ I then
11. swap(A[i],A[largest])
12. Max_Heap(A,largest)
13. }
```

The above function will arrange the elements into a max heap.

3.5.1 Building a Heap

To convert an array A[1 … n] where n=length[A] into a max heap, we have to use the above-said algorithm Max_Heap.

In a complete Binary Tree of n nodes out of which (n/2) are non-leaf nodes and (n/2)+1 are the leaf nodes.

Therefore, while building a heap from an array A[1 … n] of n elements the subarray A[((n/2)+1) … n] are all leaves of the tree.

Therefore, each element of the subarray A[((n/2)+1) … n] is a one-element heap to begin with

Algorithm Build_Heap(A)

```
1. {
2. heapsize[A] ← length[A]
3. for i ← length[A]/2 to 1 by -1
4. do Max_Heap(A,i)
5. }
```

Algorithm Heap_sort(A)

```
1. {
2. Build_Heap(A)
3. for i ← length[A] to 2 by -1
4. do swap(A[1], A[i])
```

```
5. heapsize[A] ← heapsize[A]-1
6. Max_Heap(A,1)
7. }
```

Example:

Construct a HEAP and Sort them. By using Heap Sort Algorithm

12 32 54 10 5 89 76 51 09

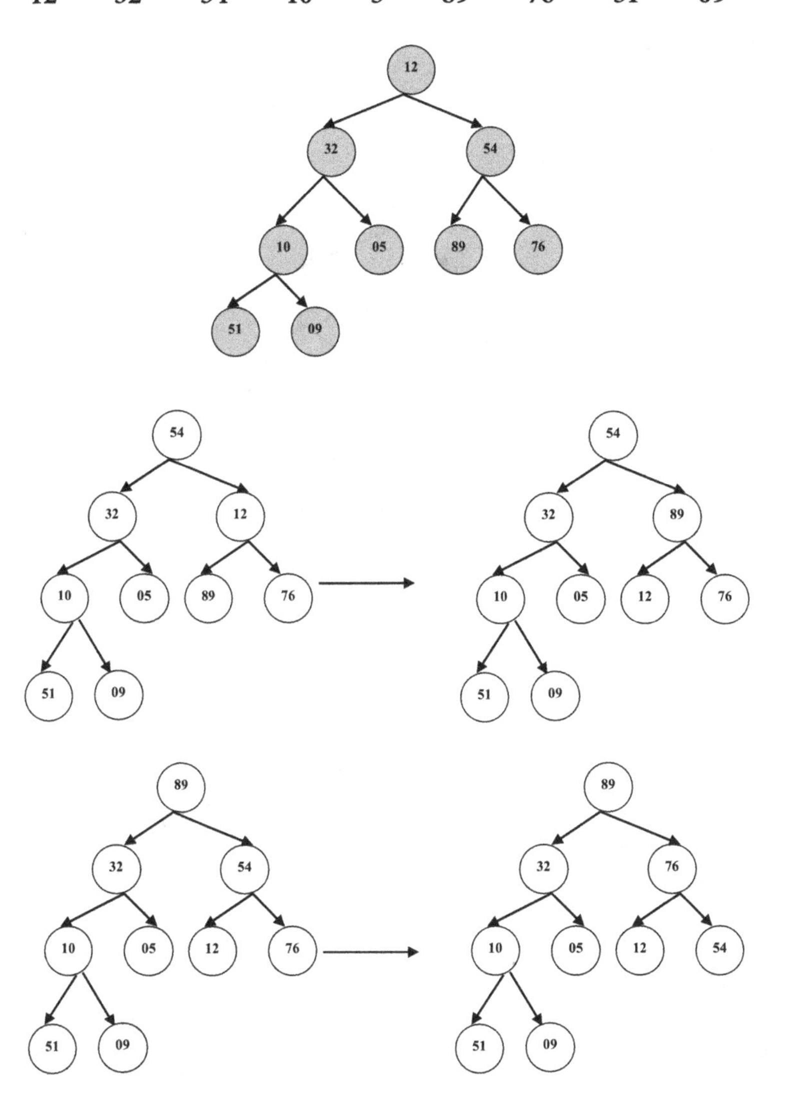

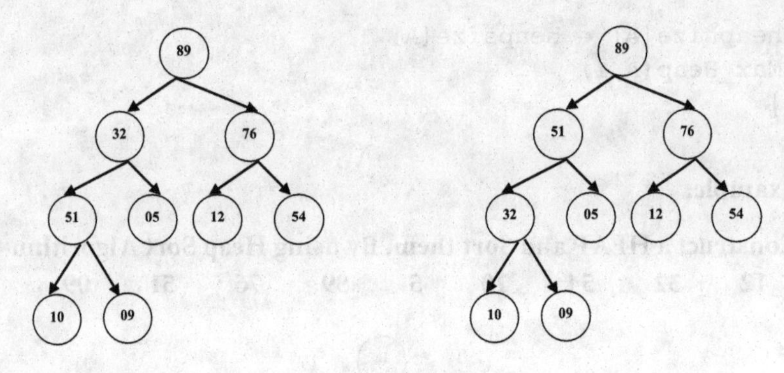

Delete 89

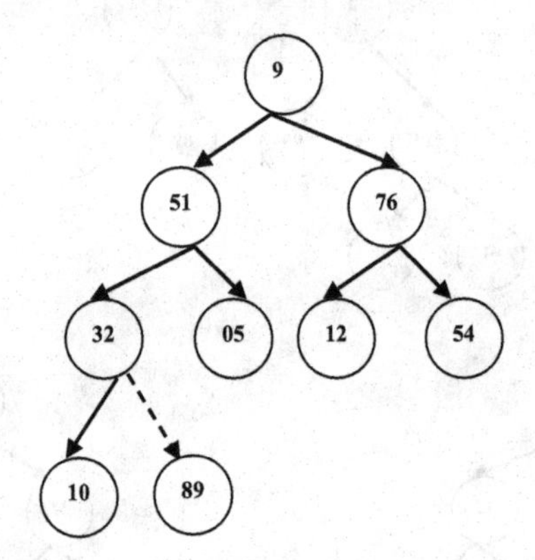

Again, build the heap tree.

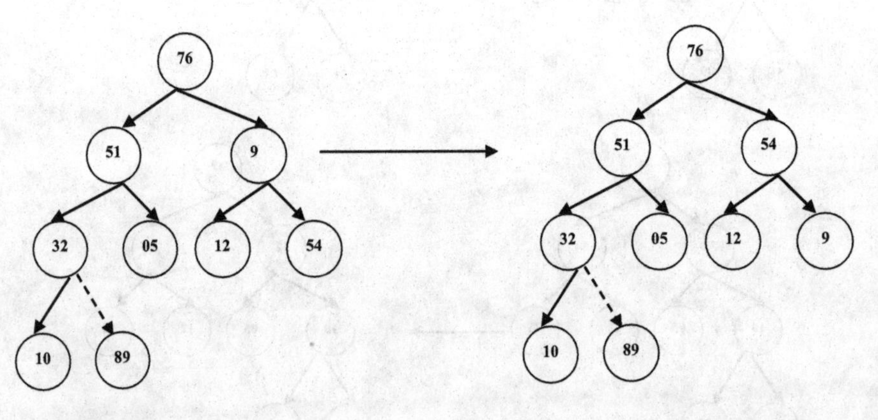

Delete 76

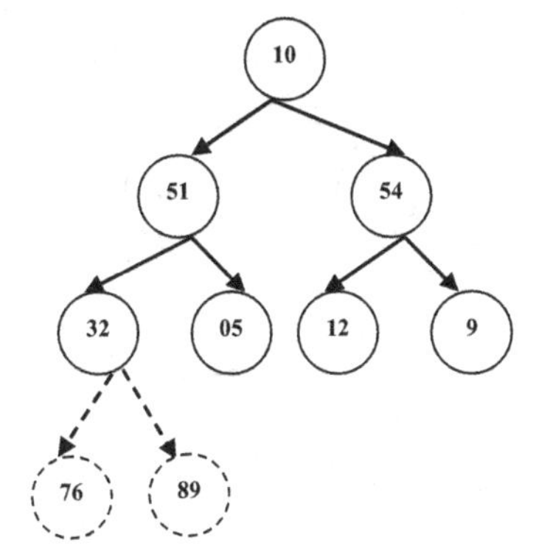

Again, build the heap tree

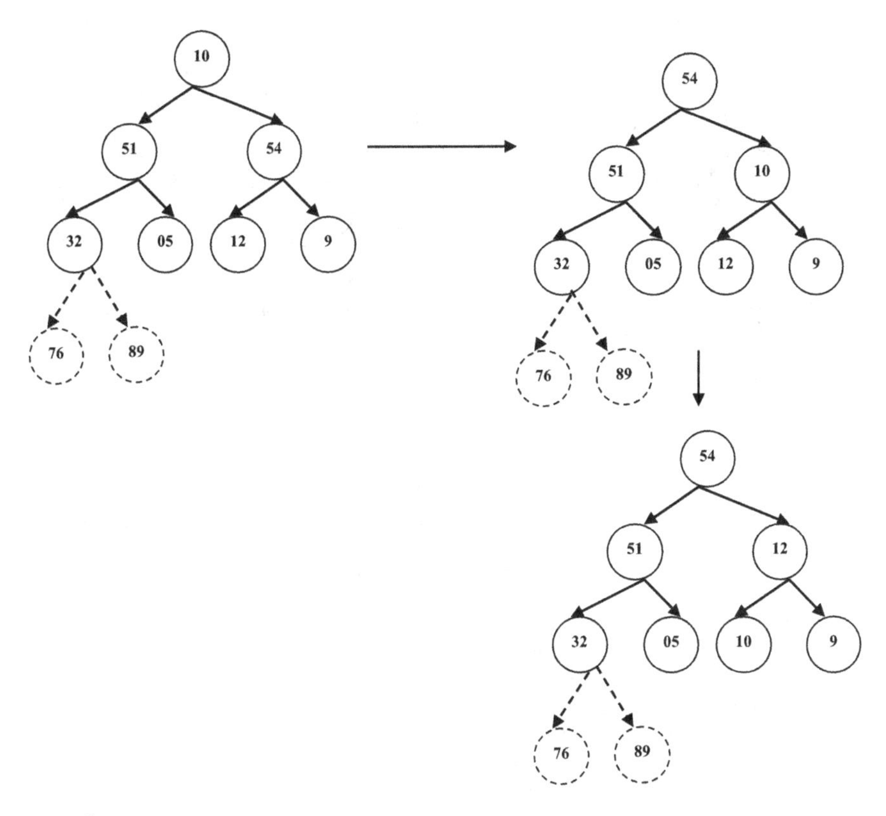

Delete 54

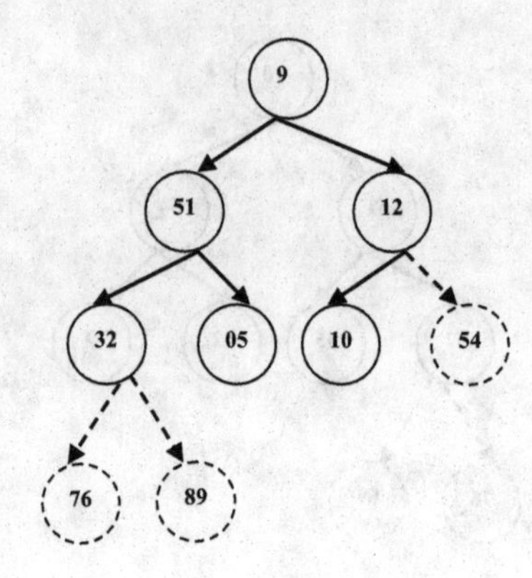

Build the heap

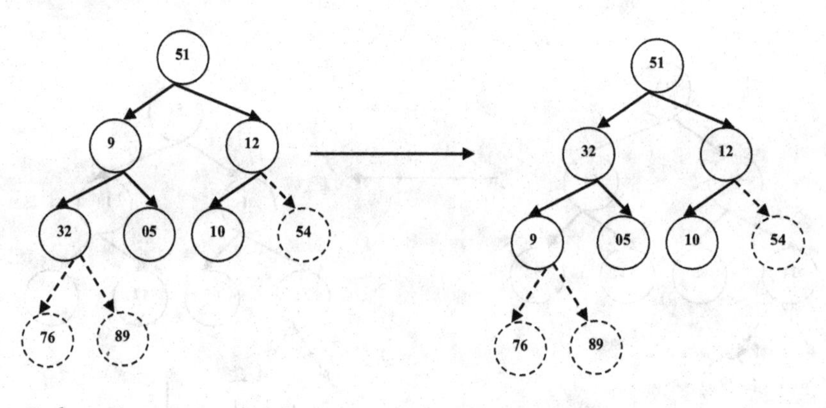

Delete 51

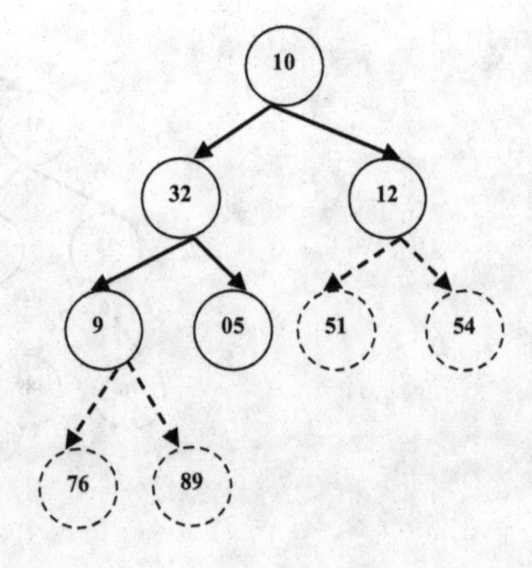

Build the heap

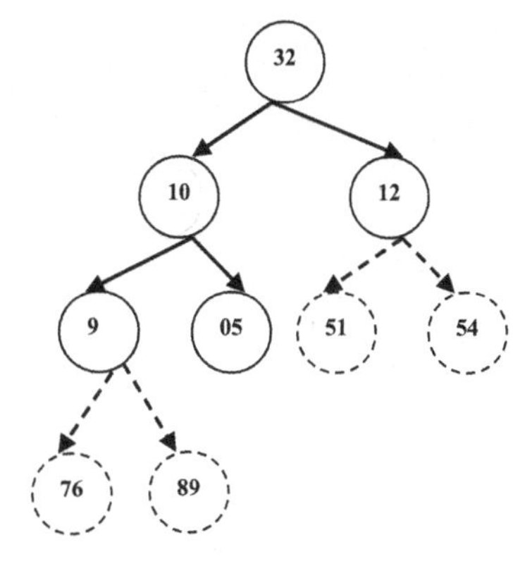

Delete 32

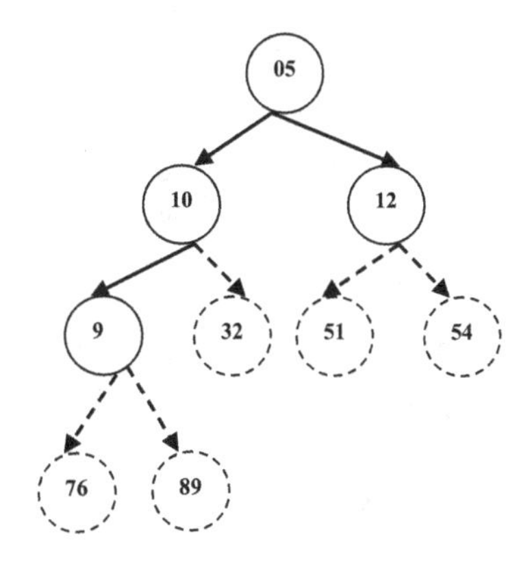

Build the heap

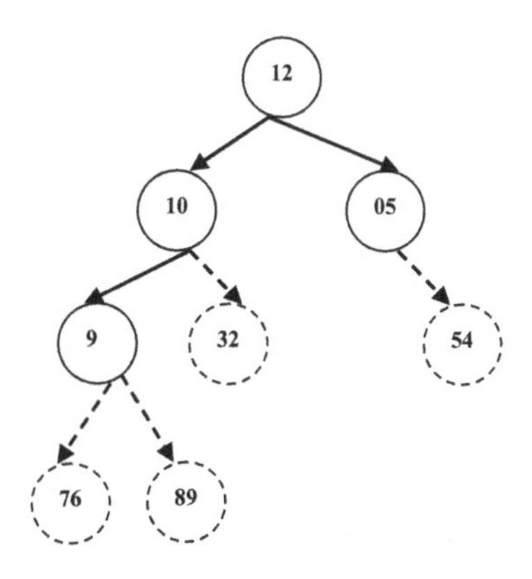

Delete 12

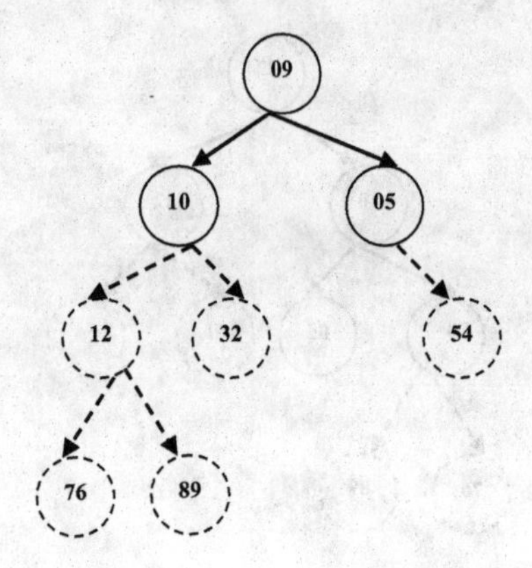

Build the heap

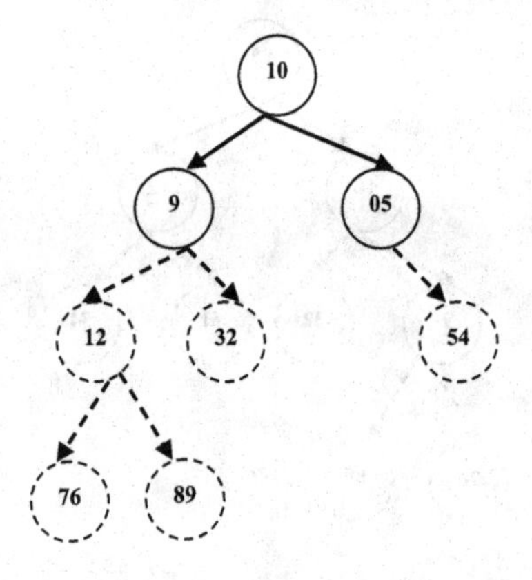

Delete 10

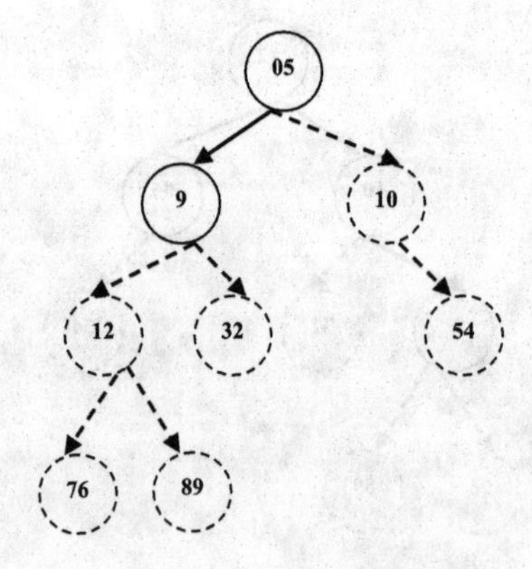

Build the heap

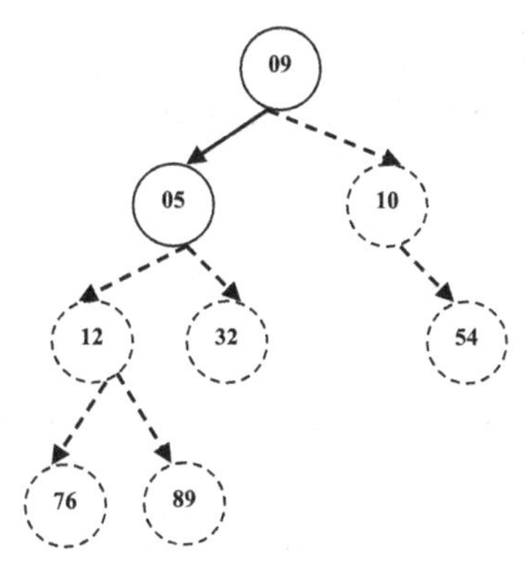

Delete 9

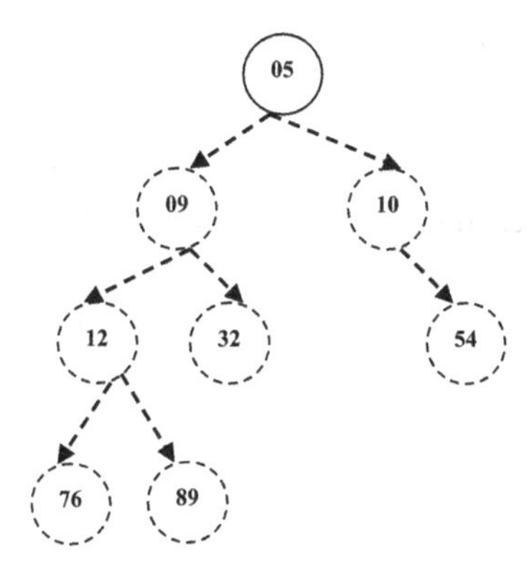

Delete 5

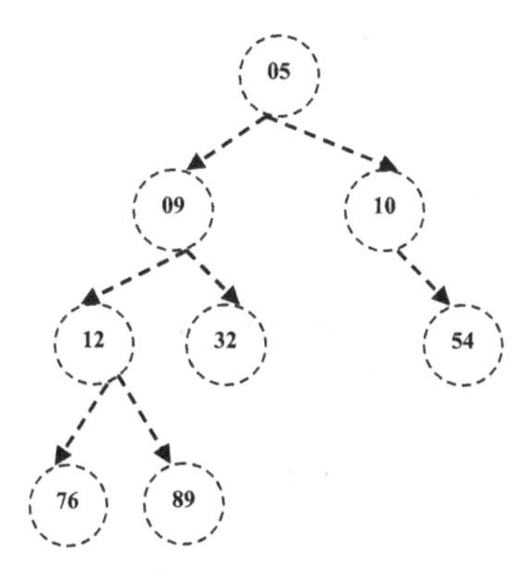

Now perform the level order traversal to get the sorted list.

i. 5, 9, 10, 12, 32, 54, 76, 89

3.6 PRIORITY QUEUE

Priority queues are a kind of queue in which the elements are dequeued in priority order.

- They are a mutable data abstraction: enqueues and dequeues are destructive.
- Each element has a **priority**, an element of a totally ordered set (usually a number)
- More important things come out *first*, even if they were added later.
- Our convention: smaller number=higher priority
- There is no (fast) operation to find out whether an arbitrary element is in the queue.
- Useful for event-based simulators (with priority=simulated time), real-time games, searching, routing and compression via Huffman coding.

Depending on the heaps the priority queue are also of two types such as

Min Priority Queue

Max Priority Queue

The set of operations for Max Priority Queue are

- **Insert(A,N)**: Inserts an element N into A
- **Maximum(A,X)**: Finds the X from A where X is the Maximum
- **Extract_Max(A)**: Remove and returns the element of S with the largest key
- **Increase_Key(A,x,k)**: Increases the value of element x's to the new value k which is assumed to be at least as large as x's current key value.

The set of operations for Min Priority Queue are

- Insert
- Minimum

- Extract_Min
- Decrease_Key

The most important application of Max Priority Queue is to schedule jobs on a shared computer. The Max Priority queue keeps track of the jobs to be performed and their relative priorities. When a job is finished or interrupted, the highest priority job is selected from those pending using Extract-Max. A new job can be added to the queue at any time by using Insert.

EXTRACT_MAX operation returns the largest element. Therefore, finding the largest element from an unordered list takes $\theta(n)$ times. An alternative is to use an ordered linear list. The elements are in nondecreasing order if a sequential representation is used.

The Extract-Max operation takes $\theta(1)$, and the Insert time is $O(n)$. When max heap is used, both Extract-max and insert can be performed in $O(\log n)$ time.

Algorithm MAXIMUM(A,X)

```
1. return A[1]
```

Algorithm EXTRACT-MAX(A)

```
1. If heapsize[A]<1
2. then Write : "Heap Underflow"
3. Max ← A[1]
4. A[1] ← A[heapsize[A]]
5. heapsize[A] ← heapsize[A]-1
6. Max_Heap(A,1)
7. return Max
```

Algorithm INCREASE-KEY(A,i,key))

```
1. if key <A[i]
2. then write "key is smaller than the current key"
3. A[i] ← key
4. while i > 1 and A[PARENT(i)]< A[i]
5. do exchange  A[i] ↔ A[PARENT(i)]
6. i ← Parent(i)
```

Algorithm INSERT(A,key)

```
1. heapsize[A] ← heapsize[A] + 1
2. A[heapsize[A]] ← -∞
3. INCREASE-KEY(A, heapsize[A],key)
```

The running time of INSERT on an n-element heap is O(lgn).

A heap can support any priority queue operation on a set of size n in O(lgn) time.

3.6.1 Operations for Min Priority Queue

Algorithm MINIMUM(A,X)
```
1. return A[1]
```

Algorithm EXTRACT-MIN(A)
```
1. If heapsize[A]<1
2. then Write : "Heap Underflow"
3. Max ← A[1]
4. A[1] ← A[heapsize[A]]
5. heapsize[A] ← heapsize[A]-1
6. Max_Heap(A,1)
7. return Min
```

Algorithm DECREASE-KEY(A,i,key))
```
1. if key >A[i]
2. then write "key is greater than the current key"
3. A[i] ← key
4. while i > 1 and A[PARENT(i)]> A[i]
5. do exchange  A[i] ↔ A[PARENT(i)]
6. i ← Parent(i)
```

Algorithm INSERT(A,key)
```
1. heapsize[A] ← heapsize[A] + 1
2. A[heapsize[A]] ← -∞
3. DECREASE-KEY(A, heapsize[A],key)
```

3.7 LOWER BOUND FOR SORTING

Sorting algorithms can be depicted as trees. The one in the following figure sorts an array of three elements, namely, a1, a2 and a3.

It starts by comparing a1 to a2 and, if the first is larger, compares it with a3; otherwise, it compares a2 and a3, and so on. Eventually, we end up at a leaf, and this leaf is labeled with the true order of the three elements as a permutation of 1, 2 and 3. For example, if a2 < a1 < a3, then we get the leaf labeled 2 1 3.

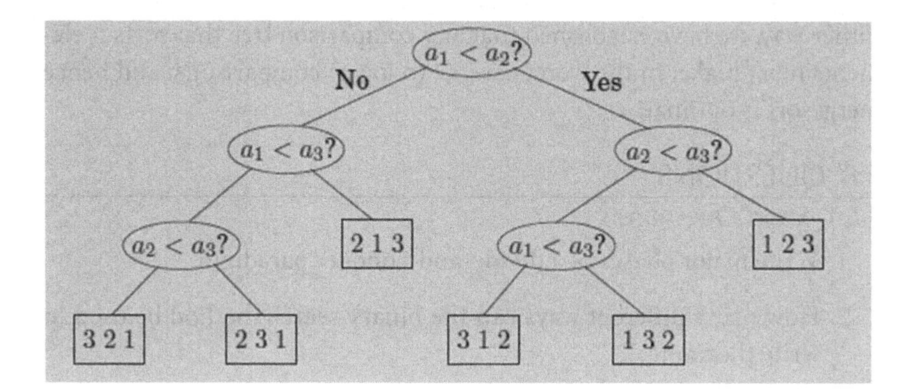

The *depth* of the tree: the number of comparisons on the longest path from root to leaf, in case 3, is exactly the worst-case time complexity of the algorithm.

This way of looking at sorting algorithms is useful because it allows one to argue that *mergesort is optimal*, in the sense that Ω (n log n) comparisons are necessary for sorting n elements.

Here is the argument: Consider any such tree that sorts an array of n elements. Each of its leaves is labeled by a permutation of {1, 2, …, n}. *Every* permutation must appear as the label of a leaf. The reason is simple: if a particular permutation is missing, what happens if we feed the algorithm an input ordered according to this same permutation? Since there are n! permutations of n elements, it follows that the tree has at least n! leaves.

We are almost done: This is a binary tree, and we argued that it has at least n! leaves. Recall now that a binary tree of depth d has at most 2^d leaves (proof: an easy induction on d). Therefore, the depth of our tree, and the complexity of our algorithm, must be at least log(n!).

It is well known that log(n!) >= c (n log n) for some c>0. There are many ways to see this.

The easiest is to notice that $n! > (n/2)^{(n/2)}$ because $n! = 1 \cdot 2 \cdot 3 \ldots n$ contains at least n/2 factors larger than n/2 and to then take logs of both sides. Another is to recall Stirling's formula:

$$n! = \sqrt{\pi\left(2n + \frac{1}{3}\right)} \cdot n^n \cdot e^{-n}.$$

Either way, we have established that any comparison tree that sorts n elements must make, in the worst case, Ω (n log n) comparisons, and hence mergesort is optimal!

3.8 QUESTIONS

3.8.1 Short Questions

1. Write major phases of a divide-and-conquer paradigm.
2. How many different ways can the binary search method be used to write the names?
3. Write the worst case of a binary tree.
4. Difference between merge sort and quick sort.
5. Define binary search and time complexity of both cases.
6. Define how merge sort works.
7. How does a divide-and-conquer method work in quick sort?
8. Discuss the worst case of quick sort.
9. Define heap sort and how it works.
10. What is a priority queue?
11. When best-case and worst-case partitioning will occur in quick sort?

3.8.2 Long Questions

1. Sort the elements 5, 14, 2, 9, 21, 34, 17, 19, 1 and 44 by using quicksort with the algorithm.
2. Construct a HEAP and sort them by using heap sort algorithm 12, 32, 54, 10, 5, 89, 76, 51 and 09.
3. Write down the algorithm for merge sort and merge procedure. Prove that the running time of merge sort is Θ (nlgn).
4. Write the algorithm for priority queue for MAX and MIN.
5. Write the algorithm for merge sort and given an example. How merge sort works?

6. Write binary search algorithm.
7. Explain the procedure of merge sort with the algorithm.
8. Explain the procedure of quick sort with the algorithm.
9. What is heap sort? And proof a heap T storing n elements has height $\log(n+1)$.
10. Write algorithm for max_heap.

CHAPTER 4

Dynamic Programming

4.1 DYNAMIC PROGRAMMING

Dynamic programming is a very powerful technique to solve a particular class of problems. It demands very elegant formulation of the approach and simple thinking, and the coding part is very easy. The idea is very simple: if you have solved a problem with the given input, then save the result for future reference, so as to avoid solving the same problem again. If the given problem can be broken up into smaller subproblems and these smaller subproblems are in turn divided into still-smaller ones, and in this process, if you observe some overlapping subproblems, then it's a big hint for DP. In addition, the optimal solutions to the subproblems contribute to the optimal solution of the given problem (referred to as the optimal substructure property).

There are two ways of doing this:

1. **Top-Down**: Start solving the given problem by breaking it down. If you see that the problem has been solved already, then just return the saved answer. If it has not been solved, solve it and save the answer. This is usually easy to think of and very intuitive. This is referred to as **Memoization**.
2. **Bottom-Up**: Analyze the problem and see the order in which the subproblems are solved and start solving from the trivial subproblem, up towards the given problem. In this process, it is guaranteed that the subproblems are solved before solving the problem. This is referred to as **Dynamic programming**.

DOI: 10.1201/9781003093886-4

Dynamic programming is often used to solve optimization problems. In these cases, the solution corresponds to an objective function whose value needs to be optimal (e.g., maximal or minimal). In general, it is sufficient to produce one optimal solution, even though there may be many optimal solutions for a given problem.

The principle of optimality states that an optimal sequence of decisions has the property that whatever the initial state and decision are, the remaining decisions must constitute an optimal decision sequence with regard to the state resulting from the first decision.

4.2 DEVELOPING DYNAMIC PROGRAMMING ALGORITHMS

Three basic elements characterize a dynamic programming algorithm:

1. **Substructure**: Decompose the given problem into smaller (and hopefully simpler) subproblems. Express the solution to the original problem in terms of solutions to smaller problems. Note that unlike divide-and-conquer problems, it is not usually sufficient to consider one decomposition, but many different ones.

2. **Table Structure**: After solving the subproblems, store the answers (results) to the subproblems in a table. This is done because (typically) subproblem solutions are reused many times, and we do not want to repeatedly solve the same problem.

3. **Bottom-Up Computation**: Using table (or something), combine solutions of smaller subproblems to solve larger subproblems and eventually solve the complete problem. The idea of bottom-up computation is as follows:

Bottom-up means

i. Start with the smallest subproblems.

ii. Combining their solutions obtain the solutions to subproblems of increasing size.

iii. Until solve the original problem.

Once we decided that we are going to attack the given problem with a dynamic programming technique, the most important step is the **formulation of the problem**. In other words, the most important question in designing a dynamic programming solution to a problem is how to set up the subproblem structure.

4.2.1 Optimal Substructure

Show that a solution to a problem consists of making a choice, which leaves one or subproblems to solve. Now suppose that you are given this last choice to an optimal solution. Given this choice, determine which subproblems arise and how to characterize the resulting space of subproblems. Show that the solutions to the subproblems used within the optimal solution must themselves be optimal (optimality principle). You usually use cut-and-paste:

- Suppose that one of the subproblems is not optimal.
- Cut it out.
- Paste in an optimal solution.
- Get a better solution to the original problem. Contradicts optimality of problem solution.

That was an optimal substructure.

We have used the "Optimality Principle" a couple of times. Now a word about this beast: the optimal solution to the problem contains within it optimal solutions to subproblems. This is sometimes called the principle of optimality.

4.2.2 The Principle of Optimality

Dynamic programming relies on a principle of optimality. This principle states that in an optimal sequence of decisions or choices, each subsequence must also be optimal. For example, in the matrix chain multiplication problem, not only the value we are interested in is optimal but all the other entries in the table also represent optimal. The principle can be related as follows: the optimal solution to a problem is a combination of optimal solutions to some of its subproblems. The difficulty in turning the principle of optimality into an algorithm is that it is not usually obvious which subproblems are relevant to the problem under consideration.

Now the question is how to characterize the space of subproblems?

- Keep the space as simple as possible.
- Expand it as necessary.

Optimal substructure varies across problem domains:

1. How *many subproblems* are used in an optimal solution?
2. How *many choices* are in determining which subproblem(s) to use.

Overall, four steps are required to develop the dynamic programming algorithms such as

- Characterize the structure of an optimal solution.
- Recursively define the value of an optimal solution.
- Compute the value of an optimal solution in a bottom-up fashion.
- Construct an optimal solution from the computed information.

The first three components are the basis of a dynamic programming solution to a problem. Step 4 can be omitted if only the value of an optimal solution is required

Some implementations of dynamic programming algorithms are

- Assembly line scheduling
- Knapsack problem
- Longest common subsequence problem
- Matrix chain multiplication
- Mathematical optimization problems.

4.3 MATRIX CHAIN MULTIPLICATION

Let A be a (pxq) matrix and B be a (qxr) matrix. Then, we can multiply

A(pxq) * B(q * r) = C(pxr), where the elements of C(pxr) are defined as:

$$Cij = \sum_{k=1}^{q} AikBkj$$

This algorithm takes p * q * r multiplications to compute $A_{pxq} * B_{qxr}$

4.3.1 Chains of Matrices

Consider A1 … An. We can compute this product if the number of columns in Ai is equal to the number of rows in Ai+1 (Cols[Ai]=Rows[Ai+1]) for every

$1 \leq i \leq n - 1$.

4.3.1.1 How to Order Multiplications?

Notice that matrix multiplication is associative i.e./$A \cdot (B \cdot C) = (A \cdot B) \cdot C$. This results in many possible parenthesizations (i.e., orderings of matrix multiplications).

A product of matrices is fully parenthesized if it is either a single matrix or the product of two fully parenthesized matrix products.

For example, A1 * A2 * A3 * A4 can be fully parenthesized as

$(A1(A2(A3 \cdot A4)))$

$(A1 \cdot ((A2 \cdot A3) \cdot A4))$

$((A1 \cdot A2) \cdot (A3 \cdot A4))$

$(((A1 \cdot A2) \cdot A3) \cdot A4)$

$((A1 \cdot (A2 \cdot A3)) \cdot A4)$

4.3.1.1.1 The Parentherisation Is Important

Different parenthesizations of A1 … An may lead to different total numbers of scalar multiplications.

For example Let A1 be a 10 * 100 matrix

A2 be a 100 * 5 matrix

A3 be a 5 * 50 matrix then

$Cost[(A1 \cdot A2) \cdot A3] = (10 * 100 * 5) + (10 * 5 * 50)$

$= 7{,}500$ multiplications.

And

$Cost[A1 \cdot (A2 \cdot A3)] = (100 * 5 * 50) + (10 * 100 * 50) = 75{,}000$ Multiplications.

In comparison with the above two, the first way will be optimal.

Algorithm MatrixMul(A,B)

```
 1. if col[A] ≠ row[B]
 2. Then "Incompatible dimensions"
 3. else
 4. for i ← 1 to row[A]
 5. do for j ← 1 to col[B]
 6. do C[i,j] ← 0
 7. for k ← 1 to col[A]
 8. do C[i,j] ← C[i,j] + A[i,k] * B[k,j]
 9. [end loop]
10. [end loop]
11. [end loop]
12. return C
```

CASE 1: Determine the Structure of An Optimal Solution (Optimal Parenthesization)

The first step in the dynamic programming paradigm is to find the optimal sub-structure and then use it to construct an optimal solution to the problem from optimal solutions to subproblems. For the matrix chain multiplication, we can perform the steps as:

Let's consider the matrices $A_i \ldots A_j$ where $i \leq j$.

Now if the problem is nontrivial, i.e., $i<j$ then any parenthesization of the product $A_i, A_{i+1} \ldots A_j$ must split the product between A_k and A_{k+1} for some integer K in the range $i \leq k<j$. That is, for some k, we first compute $A_i \ldots k$ and $A_{k+1} \ldots j$ and then multiply them to get the final result.

The cost of this parenthesization is thus the cost of computing the matrix $A_i \ldots k$ plus the cost of computing $A_{k+1} \ldots j$, plus the cost of multiplying them together.

The optimal substructure of this subproblem can be discussed as:

Suppose that an optimal parenthesization of $A_i, A_{i+1}, \ldots A_j$ splits the product between A_k and A_{k+1}. Then the parenthesization of the prefix subchain $A_i, A_{i+1}, \ldots A_k$ within this optimal parenthesization of $A_i, A_{i+1} \ldots A_j$ must be an optimal parenthesization of $A_i, A_{i+1}, \ldots A_k$. If there were a less costly way to parenthesize $A_i, A_{i+1}, \ldots A_k$, substituting that parenthesization in the optimal parenthesization of Ai,Ai+1,….Aj would produce another parenthesization of $A_i, A_{i+1}, \ldots A_j$ whose cost was lower than the optimum: a contradiction. A similar observation holds for the parenthesization of the sub chain $A_{k+1}, A_{k+2}, \ldots A_j$ in the optimal parenthesization of $A_i, A_{i+1}, \ldots A_j$: it must be an optimal parenthesization of $A_{k+1}, A_{k+2}, \ldots A_j$.

Now we use our optimal substructure to show that we can construct an optimal solution to the problem from optimal solutions to subproblems.

CASE 2: A Recursive Solution

We can define m[i, j] as:

If i=j, then the problem is trivial, then the chain consists of just one matrix i.e./A_i … i =A_i so that no scalar multiplications are necessary to compute the product.

Thus m[i, j]=0 for i=1, 2, …, n

If i<j then we have to use the optimal subproblem from step-1

Let us assume that the optimal parenthesization splits the product A_i, A_{i+1}, … A_j between A_k and A_{k+1}, where i≤k<j. Then m[i, j] is equal to the minimum cost for computing the sub-products A_i … k and A_{k+1} … j, plus the cost of multiplying these two matrices together.

Recalling that each matrix A_i is Pi−1 X Pi, we observe that the matrix product A_i … k_{Ak}+1 … j takes Pi−1PkPj scalar multiplications so mathematically it can be represented as

$$m[i,j]=m[i,k]+m[k+1,j]+Pi-1PkPj$$

Since our requirement is the optimal solution, so we can define the recursive solution as

$$m[i,j]= \begin{cases} 0 & \text{if } i=j \\ \min\{m[i,k]+m[k+1,j]+Pi-1PkPj\} & \forall i \le k < j \text{ and } i<j \end{cases}$$

The m[i, j] values give the costs of optimal solutions to subproblems.

CASE-3 Computing the Optimal Cost

A recursive algorithm may encounter each subproblem many times in different branches of the recursion tree. This property of the **Overlapping Subproblem** is the second hallmark of the applicability of dynamic programming (the first hallmark is optimal substructure). Instead of computing the solution to recurrence recursively, we perform the third step to the dynamic programming paradigm and compute the optimal cost by using a tabular, bottom-up approach.

This procedure assumes that matrix Ai has dimensions Pi−1 X Pi for i=1, 2, … n.

The input is a sequence P=<Po, P1, … Pn>, where length[P]=n+1.

Here an auxiliary table m[1 … n, 1 … n] is being used for storing the m[i, j] costs and another auxiliary table S[1 … n, 1 … n] is used which stores that which index of k achieved the optimal cost in computing m[i, j]. Finally, we will use table S to construct an optimal solution.

Algorithm MatrixChain_Order(P)

```
 1. n ← length[p]-1
 2. for i ← 1 to n
 3. do m[i,i] ← 0
 4. for len ← 2 to n
 5. do for i ← 1 to n-len +1
 6. do j←i + len -1
 7. m[i,j] ← ∞
 8. for k ← i to j-1
 9. do q ← m[i,k] + m[k+1,j] + Pi-1PkPj
10. if q < m[i,j]
11. then m[i,j] ← q
12. s[i,j] ← k
13. return m and s
```

The above algorithm first computes m[i, i] ← 0 for i=1, 2, … n in line 2–3. It then uses recurrences to compute m[i, I+1] for i ← 1, 2, … n−1 (the minimum costs for chains of length len =2) during the first execution of the loop in the lines 4–12. In the second time through the loop, it computes m[i, i+2] for i=1, 2, … n−2 (the minimum costs for chains of length len =3) and so on.

At each step, the m[i, j] cost computed in lines 9–12 depends only on table entries m[i, k] and m[k+1, j] already computed.

CASE 4: Constructing an Optimal Solution

No doubt, the above algorithm determines the optimal number of scalar multiplications needed to compute a matrix chain product, it does not directly show how to multiply the matrices. It is too easy to calculate the optimal solution by taking S[1 … n, 1 … n]. Each entry S[i, j] records the value of k such that the optimal parenthesization of $A_i \ldots A_j$ splits the product between A_k and A_{k+1}. Thus, we know that the final matrix multiplication in computing $A_1 \ldots n$ optimally is

$A_1 \ldots s[1, n] A_s [1, n] + 1 \ldots n$.

Algorithm Optimal_Parenthesis(A,S,i,j)

```
1. if i = j
2. then print "A"
3. else
4. x ← Optimal_Parenthesis(A,S,i,S[i,j])
5. y ← Optimal_Parenthesis(A,S,S[i,j]+1,j)
6. return MatrixMul(x,y)
```

Example-1:

Consider the following matrix and its dimensions as

M_1	M_2	M_3	M_4	M_5	M_6
30×35	35×15	15×5	5×10	10×20	20×25

Find the matrix chain multiplication $M_{2 \ldots 5}$

First, we have to find the vector d as: d=(30, 35, 15, 5, 10, 20, 25)

Where

$d_0=30$

$d_1=35$

$d_2=15$

$d_3=05$

$d_4=10$

I →

j ↓

	1	2	3	4	5	6
6	m16	m26	m36	m46	m56	m66
5	m15	m25	m35	m45	m55	
4	m14	m24	m34	m44		
3	m13	m23	m33			
2	m12	m22				
1	m11					

For diagonal **s=0**

$d_5=20$

$d_6=25$

$m_{11}=0$	$m_{22}=0$	$m_{33}=0$	$m_{44}=0$	$m_{55}=0$	$m_{66}=0$

For diagonal **s = 1**

$$m_{i,i+s} = d_{i-1}d_i d_{i+s},\ i=1,2,\ldots n-1$$

$$m_{12} = d_0 d_1 d_2 = 30\times35\times15 = 15{,}750 \quad i=1$$

$$m_{23} = d_1 d_2 d_3 = 35\times15\times5 = 2{,}625 \quad i=2$$

$$m_{34} = d_2 d_3 d_4 = 15\times5\times10 = 750 \quad i=3$$

$$m_{45} = d_3 d_4 d_5 = 5\times10\times20 = 1{,}000 \quad i=4$$

$$m_{56} = d_4 d_5 d_6 = 10\times20\times25 = 5{,}000 \quad i=5$$

For diagonal **s = 2**

$$mi, i+s = \min\{m_{ik} + m_{k+1,i+s} + d_{i-1}d_k d_{i+s}\}$$

$$i <= k < i+s \quad \text{for} \quad I=1,2,\ldots n-s$$

$$m_{13} = \min = \begin{cases} m_{11} + m_{23} + d_0 d_1 d_3 & k=1 \\ m_{12} + m_{33} + d_0 d_2 d_3 & k=2 \end{cases}$$

$$\min\begin{cases} 0 + 2{,}625 + 30\times35\times5 = \mathbf{7{,}875} \\ 15{,}750 + 0 + 30\times15\times5 = 18{,}000 \end{cases}$$

$$m_{24} = \min\begin{cases} m_{22} + m_{34} + d_1 d_2 d_4 & k=2 \\ m_{23} + m_{44} + d_1 d_3 d_4 & k=3 \end{cases}$$

$$\min\begin{cases} 0 + 750 + 35\times15\times10 = 6{,}000 \\ 2{,}625 + 0 + 35\times5\times10 = \mathbf{4{,}375} \end{cases}$$

$$m_{35} = \min\begin{cases} m_{33} + m_{45} + d_2 d_3 d_5 & k=3 \\ m_{34} + m_{55} + d_2 d_4 d_5 & k=4 \end{cases}$$

$$\min\begin{cases}0+1{,}000+15\times5\times20=\mathbf{2{,}500}\\750+0+15\times10\times20=3{,}750\end{cases}$$

$$m_{46}=\min\begin{cases}m_{44}+m_{56}+d_3d_4d_6 & k=4\\m_{45}+m_{66}+d_3d_5d_6 & k=5\end{cases}$$

$$\min\begin{cases}0+5{,}000+5\times10\times25=6{,}250\\1{,}000+0+5\times20\times25=\mathbf{3{,}500}\end{cases}$$

For diagonal $\mathbf{s=3}$

$$m_{14}=\min\begin{cases}m_{12}+m_{34}+d_0d_2d_4\\m_{11}+m_{24}+d_0d_1d_4\\m_{13}+m_{44}+d_0d_3d_4\end{cases}$$

$$\min\begin{cases}15{,}750+750+30\times15\times10=21{,}000\\0+4{,}375+30\times35\times10=14{,}875\\7{,}875+0+30\times5\times10=\mathbf{9{,}375}\end{cases}$$

$$m_{25}=\min\begin{cases}m_{23}+m_{45}+d_1d_3d_5\\m_{22}+m_{35}+d_1d_2d_5\\m_{24}+m_{55}+d_1d_4d_5\end{cases}$$

$$\min\begin{cases}2{,}625+1{,}000+35\times5\times20=\mathbf{7{,}125}\\0+2{,}500+35\times15\times20=13{,}000\\4{,}375+0+35\times10\times20=11{,}375\end{cases}$$

$$m_{36}=\min\begin{cases}m_{34}+m_{56}+d_2d_4d_6\\m_{33}+m_{46}+d_2d_3d_6\\m_{35}+m_{66}+d_2d_5d_6\end{cases}$$

$$\min\begin{cases}750+5{,}000+15\times10\times25=13{,}775\\0+3{,}500+15\times5\times25=\mathbf{5{,}375}\\2{,}500+0+15\times20\times25=10{,}000\end{cases}$$

For diagonal **s = 4**

$$m_{15}=\min\begin{cases}m_{11}+m_{25}+d_0d_1d_5\\m_{14}+m_{55}+d_0d_4d_5\\m_{12}+m_{35}+d_0d_2d_5\\m_{13}+m_{45}+d_0d_3d_5\end{cases}$$

$$\min\begin{cases}0+7{,}125+30\times35\times20=28{,}125\\9{,}375+0+30\times10\times20=15{,}375\\15{,}750+2{,}500+30\times15\times20=27{,}250\\7{,}875+1{,}000+30\times5\times20=\mathbf{11{,}875}\end{cases}$$

$$m_{26}=\min\begin{cases}m_{22}+m_{36}+d_1d_2d_6\\m_{23}+m_{46}+d_1d_3d_6\\m_{24}+m_{56}+d_1d_4d_6\\m_{25}+m_{66}+d_1d_5d_6\end{cases}$$

$$\min\begin{cases}0+5{,}375+35\times15\times25=18{,}500\\2{,}625+3{,}500+35\times5\times25=\mathbf{10{,}500}\\4{,}375+5{,}000+35\times10\times25=18{,}125\\7{,}125+0+35\times20\times25=24{,}625\end{cases}$$

For diagonal **s = 5**

$$m_{16} = \min \begin{cases} m_{11} + m_{26} + d_0 d_1 d_6 \\ m_{12} + m_{36} + d_0 d_2 d_6 \\ m_{13} + m_{46} + d_0 d_3 d_6 \\ m_{14} + m_{56} + d_0 d_4 d_6 \\ m_{15} + m_{66} + d_0 d_5 d_6 \end{cases}$$

$$m_{16} = \min \begin{cases} 0 + 10{,}500 + 30 \times 35 \times 25 = 36{,}750 \\ 15{,}750 + 5{,}375 + 30 \times 15 \times 25 = 32{,}375 \\ 7{,}875 + 3{,}500 + 30 \times 5 \times 25 = \mathbf{15{,}125} \\ 9{,}375 + 5{,}000 + 30 \times 10 \times 25 = 21{,}875 \\ 11{,}875 + 0 + 30 \times 20 \times 25 = 26{,}875 \end{cases}$$

i →

j ↓

	1	2	3	4	5	6
6	15125	10500	5375	3500	5000	0
5	11875	**7125**	2500	1000	0	
4	9375	4375	750	0		
3	7875	2625	0			
2	15750	0				
1	0					

S[i, j] – Table

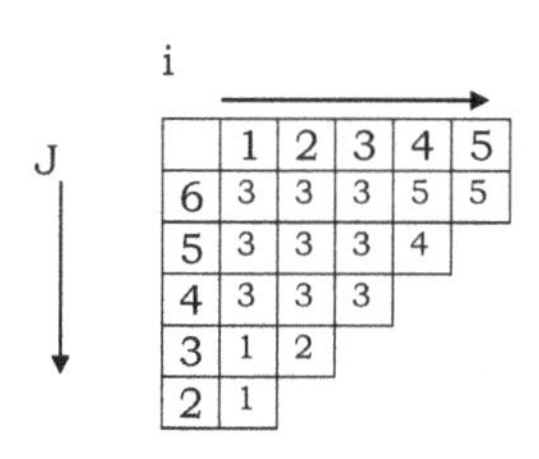

i →

J ↓

	1	2	3	4	5
6	3	3	3	5	5
5	3	3	3	4	
4	3	3	3		
3	1	2			
2	1				

M[i, j] – Table

At each stage of parenthesization, we calculate the minimum cost and is added to obtain the optimal cost. Thus, the final optimal cost for m_{25} is **7,125**.

Example:

Consider the following matrix and its dimensions as

M_1	M_2	M_3	M_4	M_5
15×10	10×25	25×15	15×5	5×30

Find the matrix chain multiplication $M_{1...5}$

First, we have to find the vector d as: d=(15, 10, 25, 15, 5, 30)

Where

$d_0=15$

$d_1=10$

$d_2=25$

$d_3=15$

$d_4=05$

$d_5=30$

I →

j ↓

	1	2	3	4	5
5	m15	m25	m35	m45	m55
4	m14	m24	m34	m44	
3	m13	m23	m33		
2	m12	m22			
1	m11				

For diagonal **s=0**

$m_{11}=0$	$m_{22}=0$	$m_{33}=0$	$m_{44}=0$	$m_{55}=0$	$m_{66}=0$

For diagonal **s=1**

$$m_{i,i+s}=d_{i-1}d_i d_{i+s},\ i=1,2,\ldots n-1$$

$$m_{12}=d_0d_1d_2=15\times10\times25=3{,}750\quad i=1$$

$$m_{23}=d_1d_2d_3=10\times25\times15=3{,}750\quad i=2$$

$$m_{34}=d_2d_3d_4=25\times15\times5=1{,}875\quad i=3$$

$$m_{45}=d_3d_4d_5=15\times5\times30=2{,}250\quad i=4$$

For diagonal **s=2**

$$mi,i+s = \min\{m_{ik} + m_{k+1,i+s} + d_{i-1}d_k d_{i+s}\}$$

$$i <= k < i+s \quad \text{for } I = 1,2,\ldots n-s$$

$$m_{13} = \min\begin{cases} m_{11} + m_{23} + d_0 d_1 d_3 & k=1 \\ m_{12} + m_{33} + d_0 d_2 d_3 & k=2 \end{cases}$$

$$\min\begin{cases} 0 + 3,750 + 15 \times 10 \times 15 = \mathbf{6,000} \\ 3,750 + 0 + 15 \times 25 \times 15 = 9,375 \end{cases}$$

$$m_{24} = \min\begin{cases} m_{22} + m_{34} + d_1 d_2 d_4 & k=2 \\ m_{23} + m_{44} + d_1 d_3 d_4 & k=3 \end{cases}$$

$$\min\begin{cases} 0 + 1,875 + 10 \times 25 \times 5 = \mathbf{3,125} \\ 3,750 + 0 + 10 \times 15 \times 5 = 4,500 \end{cases}$$

$$m_{35} = \min\begin{cases} m_{33} + m_{45} + d_2 d_3 d_5 & k=3 \\ m_{34} + m_{55} + d_2 d_4 d_5 & k=4 \end{cases}$$

$$\min\begin{cases} 0 + 2,250 + 25 \times 15 \times 30 = 13,500 \\ 1,875 + 0 + 25 \times 5 \times 30 = \mathbf{5,625} \end{cases}$$

For diagonal **s = 3**

$$m_{14} = \min\begin{cases} m_{12} + m_{34} + d_0 d_2 d_4 \\ m_{11} + m_{24} + d_0 d_1 d_4 \\ m_{13} + m_{44} + d_0 d_3 d_4 \end{cases}$$

$$\min\begin{cases} 3,750 + 1,875 + 15 \times 25 \times 5 = 7,500 \\ 0 + 3,125 + 15 \times 10 \times 5 = \mathbf{3,875} \\ 6,000 + 0 + 15 \times 15 \times 5 = 7,125 \end{cases}$$

$$m_{25} = \min \begin{cases} m_{23} + m_{45} + d_1 d_3 d_5 \\ m_{22} + m_{35} + d_1 d_2 d_5 \\ m_{24} + m_{55} + d_1 d_4 d_5 \end{cases}$$

$$\min \begin{cases} 3{,}750 + 2{,}250 + 10 \times 15 \times 30 = 10{,}500 \\ 0 + 5{,}625 + 10 \times 25 \times 30 = 13{,}125 \\ 3{,}125 + 0 + 10 \times 5 \times 30 = \mathbf{4{,}625} \end{cases}$$

For diagonal **s = 4**

$$m_{15} = \min \begin{cases} m_{11} + m_{25} + d_0 d_1 d_5 \\ m_{14} + m_{55} + d_0 d_4 d_5 \\ m_{12} + m_{35} + d_0 d_2 d_5 \\ m_{13} + m_{45} + d_0 d_3 d_5 \end{cases}$$

$$\min \begin{cases} 0 + 4{,}625 + 15 \times 10 \times 30 = 9{,}125 \\ 3{,}875 + 0 + 15 \times 5 \times 30 = \mathbf{6{,}125} \\ 3{,}750 + 5{,}625 + 15 \times 25 \times 30 = 20{,}625 \\ 6{,}000 + 2{,}250 + 15 \times 15 \times 30 = 15{,}000 \end{cases}$$

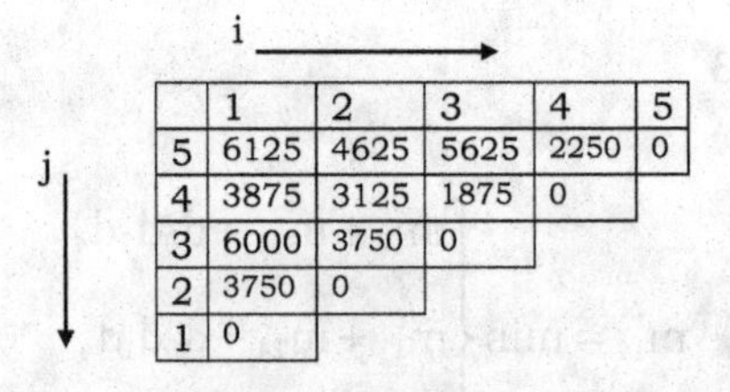

j \ i	1	2	3	4	5
5	6125	4625	5625	2250	0
4	3875	3125	1875	0	
3	6000	3750	0		
2	3750	0			
1	0				

S[i, j] – Table

J \ i	1	2	3	4
5	4	4	4	4
4	1	2	3	
3	1	2		
2	1			

M[i, j] – Table

At each stage of parenthesization, we calculate the minimum cost and is added to obtain the optimal cost. Thus, the final optimal cost for M_{15} is **6,125**.

Example:

Consider the following matrix and its dimensions as

M_1	M_2	M_3	M_4	M_5	M_6	M_7
5×12	12×10	10×20	20×5	5×30	30×10	10×5

Find the matrix chain multiplication $M_{1\ldots7}$

First, we have to find the vector d as: d=(5, 12, 10, 20, 5, 30, 10, 5)

Where

$d_0=5$

$d_1=12$

$d_2=10$

$d_3=20$

$d_4=05$

$d_5=30$

$d_6=10$

$d_7=5$

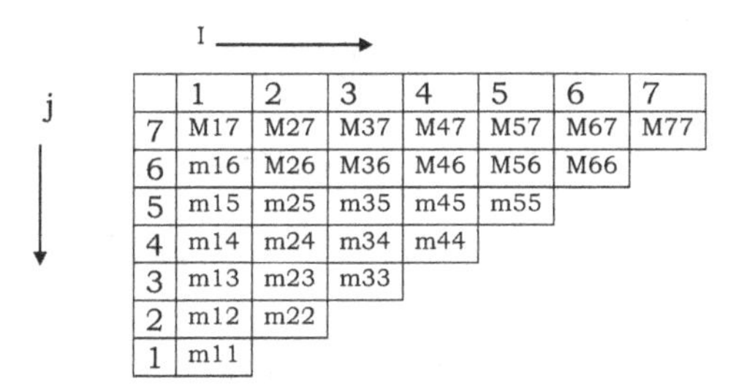

	1	2	3	4	5	6	7
7	M17	M27	M37	M47	M57	M67	M77
6	m16	M26	M36	M46	M56	M66	
5	m15	m25	m35	m45	m55		
4	m14	m24	m34	m44			
3	m13	m23	m33				
2	m12	m22					
1	m11						

For diagonal **s=0**

$m_{11}=0$	$m_{22}=0$	$m_{33}=0$	$m_{44}=0$	$m_{55}=0$	$m_{66}=0$	$m_{77}=0$

For diagonal **s=1**

$$m_{i,i+s}=d_{i-1}d_id_{i+s},\quad i=1,2,\ldots n-1$$

$$m_{12}=d_0d_1d_2=5\times12\times10=600\quad i=1$$

$$m_{23} = d_1d_2d_3 = 12\times10\times20 = 2,400 \quad i=2$$

$$m_{34} = d_2d_3d_4 = 10\times20\times5 = 1,000 \quad i=3$$

$$m_{45} = d_3d_4d_5 = 20\times5\times30 = 3,000 \quad i=4$$

$$m_{56} = d_4d_5d_6 = 5\times30\times10 = 1,500 \quad i=5$$

$$m_{67} = d_5d_6d_7 = 30\times10\times5 = 1,500 \quad i=5$$

For diagonal **s = 2**

$$mi,i+s = \min\{m_{ik} + m_{k+1,i+s} + d_{i-1}d_kd_{i+s}\}$$

$$i <= k < i+s \quad \text{for} \quad I = 1,2,\ldots n-s$$

$$m_{13} = \min\begin{cases} m_{11} + m_{23} + d_0d_1d_3 & k=1 \\ m_{12} + m_{33} + d_0d_2d_3 & k=2 \end{cases}$$

$$\min\begin{cases} 0+2,400+5\times12\times20 = 3,600 \\ 600+0+5\times10\times20 = \mathbf{1,600} \end{cases}$$

$$m_{24} = \min\begin{cases} m_{22} + m_{34} + d_1d_2d_4 & k=2 \\ m_{23} + m_{44} + d_1d_3d_4 & k=3 \end{cases}$$

$$\min\begin{cases} 0+1,000+12\times10\times5 = \mathbf{1,600} \\ 2,400+0+12\times20\times5 = 3,600 \end{cases}$$

$$m_{35} = \min\begin{cases} m_{33} + m_{45} + d_2d_3d_5 & k=3 \\ m_{34} + m_{55} + d_2d_4d_5 & k=4 \end{cases}$$

$$\min\begin{cases} 0+3,000+10\times20\times30 = 9,000 \\ 1,000+0+10\times5\times30 = \mathbf{2,500} \end{cases}$$

$$m_{46} = \min\begin{cases} m_{44} + m_{56} + d_3d_4d_6 & k=2 \\ m_{45} + m_{66} + d_3d_5d_6 & k=3 \end{cases}$$

$$\min\begin{cases}0+1{,}500+20\times5\times10=\mathbf{2{,}500}\\3{,}000+0+20\times30\times10=9{,}000\end{cases}$$

$$m_{57}=\min\begin{cases}m_{55}+m_{67}+d_4d_5d_7 & k=2\\m_{56}+m_{77}+d_4d_6d_7 & k=3\end{cases}$$

$$\min\begin{cases}0+1{,}500+05\times30\times5=2{,}250\\1{,}500+0+05\times10\times5=\mathbf{1{,}750}\end{cases}$$

For diagonal **s = 3**

$$m_{14}=\min\begin{cases}m_{12}+m_{34}+d_0d_2d_4\\m_{11}+m_{24}+d_0d_1d_4\\m_{13}+m_{44}+d_0d_3d_4\end{cases}$$

$$\min\begin{cases}600+1{,}000+5\times10\times5=\mathbf{1{,}850}\\0+1{,}600+5\times12\times5=1{,}900\\1{,}600+0+5\times20\times5=2{,}100\end{cases}$$

$$m_{25}=\min\begin{cases}m_{23}+m_{45}+d_1d_3d_5\\m_{22}+m_{35}+d_1d_2d_5\\m_{24}+m_{55}+d_1d_4d_5\end{cases}$$

$$\min\begin{cases}2{,}400+3{,}000+12\times20\times30=12{,}600\\0+2{,}500+12\times10\times30=6{,}100\\1{,}600+0+12\times5\times30=\mathbf{3{,}400}\end{cases}$$

$$m_{36}=\min\begin{cases}m_{33}+m_{46}+d_2d_3d_6\\m_{34}+m_{56}+d_2d_4d_6\\m_{35}+m_{66}+d_2d_5d_6\end{cases}$$

$$\min\begin{cases}0+2{,}500+10\times 20\times 10=4{,}500\\1{,}000+1{,}500+10\times 05\times 10=\mathbf{3{,}000}\\2{,}500+0+10\times 30\times 10=5{,}500\end{cases}$$

$$m_{47}=\min\begin{cases}m_{44}+m_{57}+d_3d_4d_7\\m_{45}+m_{67}+d_3d_5d_7\\m_{46}+m_{77}+d_3d_6d_7\end{cases}$$

$$\min\begin{cases}0+1{,}750+20\times 5\times 5=\mathbf{2{,}250}\\3{,}000+1{,}500+20\times 30\times 5=7{,}500\\2{,}500+0+20\times 10\times 5=3{,}500\end{cases}$$

For diagonal **s = 4**

$$m_{15}=\min\begin{cases}m_{11}+m_{25}+d_0d_1d_5\\m_{14}+m_{55}+d_0d_4d_5\\m_{12}+m_{35}+d_0d_2d_5\\m_{13}+m_{45}+d_0d_3d_5\end{cases}$$

$$\min\begin{cases}0+3{,}400+5\times 12\times 30=5{,}200\\1{,}850+0+5\times 5\times 30=\mathbf{2{,}600}\\600+2{,}500+5\times 10\times 30=4{,}600\\1{,}600+3{,}000+5\times 20\times 30=7{,}600\end{cases}$$

$$m_{26}=\min\begin{cases}m_{22}+m_{36}+d_1d_2d_6\\m_{23}+m_{46}+d_1d_3d_6\\m_{24}+m_{56}+d_1d_4d_6\\m_{25}+m_{66}+d_1d_5d_6\end{cases}$$

$$\min\begin{cases}0+3{,}000+12\times10\times10=4{,}200\\2{,}400+2{,}500+12\times20\times10=7{,}300\\1{,}600+1{,}500+12\times5\times10=\mathbf{3{,}700}\\3{,}400+0+12\times30\times10=7{,}000\end{cases}$$

$$m_{37}=\min\begin{cases}m_{33}+m_{47}+d_2d_3d_7\\m_{34}+m_{57}+d_2d_4d_7\\m_{35}+m_{67}+d_2d_5d_7\\m_{36}+m_{77}+d_2d_6d_7\end{cases}$$

$$\min\begin{cases}0+2{,}250+10\times20\times5=3{,}250\\1{,}000+1{,}750+10\times5\times5=\mathbf{3{,}000}\\2{,}500+1{,}500+10\times30\times5=5{,}500\\3{,}000+0+10\times10\times5=3{,}500\end{cases}$$

For diagonal $\mathbf{s=5}$

$$m_{16}=\min\begin{cases}m_{11}+m_{26}+d_0d_1d_6\\m_{12}+m_{36}+d_0d_2d_6\\m_{13}+m_{46}+d_0d_3d_6\\m_{14}+m_{56}+d_0d_4d_6\\m_{15}+m_{66}+d_0d_5d_6\end{cases}$$

$$\min\begin{cases}0+3{,}700+5\times12\times10=4{,}300\\600+3{,}000+5\times10\times10=4{,}100\\1{,}600+2{,}500+5\times20\times10=5{,}100\\1{,}850+1{,}500+5\times5\times10=\mathbf{3{,}600}\\2{,}600+0+5\times30\times10=4{,}100\end{cases}$$

$$m_{27} = \min \begin{cases} m_{22} + m_{37} + d_1 d_2 d_7 \\ m_{23} + m_{47} + d_1 d_3 d_7 \\ m_{24} + m_{57} + d_1 d_4 d_7 \\ m_{25} + m_{67} + d_1 d_5 d_7 \\ m_{26} + m_{77} + d_1 d_6 d_7 \end{cases}$$

$$\min \begin{cases} 0 + 3{,}000 + 12 \times 10 \times 5 = \mathbf{3{,}600} \\ 2{,}400 + 2{,}250 + 12 \times 20 \times 5 = 5{,}850 \\ 1{,}600 + 1{,}750 + 12 \times 5 \times 5 = 3{,}650 \\ 3{,}400 + 1{,}500 + 12 \times 30 \times 5 = 6{,}700 \\ 3{,}700 + 0 + 12 \times 10 \times 10 = 4{,}900 \end{cases}$$

For diagonal $\mathbf{s = 6}$

$$m_{17} = \min \begin{cases} m_{11} + m_{27} + d_0 d_1 d_7 \\ m_{12} + m_{37} + d_0 d_2 d_7 \\ m_{13} + m_{47} + d_0 d_3 d_7 \\ m_{14} + m_{57} + d_0 d_4 d_7 \\ m_{15} + m_{67} + d_0 d_5 d_7 \\ m_{16} + m_{77} + d_0 d_6 d_7 \end{cases}$$

$$\min \begin{cases} 0 + 3{,}600 + 5 \times 12 \times 5 = 3{,}900 \\ 600 + 3{,}000 + 5 \times 10 \times 5 = 3{,}850 \\ 1{,}600 + 2{,}250 + 5 \times 20 \times 5 = 4{,}350 \\ 1{,}850 + 1{,}750 + 5 \times 5 \times 5 = \mathbf{3{,}725} \\ 2{,}600 + 1{,}500 + 5 \times 30 \times 5 = 4{,}850 \\ 3{,}600 + 0 + 5 \times 10 \times 5 = 3{,}850 \end{cases}$$

I →

j ↓

	1	2	3	4	5	6	7
7	3725	3600	3000	2250	1750	1500	0
6	3600	3700	3000	2500	1500	0	
5	2600	3400	2500	3000	0		
4	1850	1600	1000	0			
3	1600	2400	0				
2	600	0					
1	0						

S[i, j] – Table

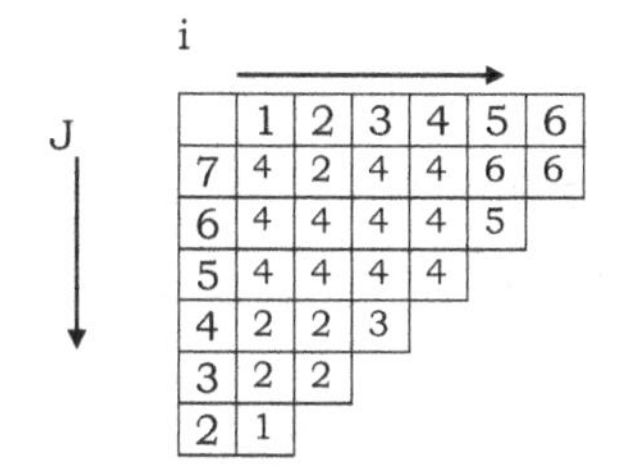

i →

J ↓

	1	2	3	4	5	6
7	4	2	4	4	6	6
6	4	4	4	4	5	
5	4	4	4	4		
4	2	2	3			
3	2	2				
2	1					

M[i, j] – Table

At each stage of parenthesization, we calculate the minimum cost and is added to obtain the optimal cost. Thus, the final optimal cost for M_{15} is **6,125.**

Example:

Let consider six matrices with the dimensions as

A_0	A_1	A_2	A_3	A_4	A_5
10×5	5×2	2×20	20×12	12×4	4×60

First, we have to find the vector d as: d=(10, 5, 2, 20, 12, 4, 60)

where

$d_0=10$

$d_1=5$

$d_2=2$

$d_3=20$

$d_4=12$

$d_5=4$

$d_6=60$

j \ I	0	1	2	3	4	5
5	A05	A15	A25	A35	A45	A55
4	A04	A14	A24	A34	A44	
3	A03	A13	A23	A33		
2	A02	A12	A22			
1	A01	A11				
0	A00					

For diagonal **s = 0**

$A_{00} = 0$	$A_{11} = 0$	$A_{22} = 0$	$A_{33} = 0$	$A_{44} = 0$	$A_{55} = 0$

For diagonal **s = 1**

$$A_{i,i+s} = d_{i-1}d_i d_{i+s}, \quad i = 1,2,\ldots n-1$$

$$A_{01} = d_0 d_1 d_2 = 10\times5\times2 = 100 \quad K = 0$$

$$A_{12} = d_1 d_2 d_3 = 5\times2\times20 = 200 \quad K = 1$$

$$A_{23} = d_2 d_3 d_4 = 2\times20\times12 = 480 \quad K = 2$$

$$A_{34} = d_3 d_4 d_5 = 20\times12\times4 = 960 \quad K = 3$$

$$A_{45} = d_4 d_5 d_6 = 12\times4\times60 = 2{,}880 \quad K = 4$$

For diagonal **s = 2**

$$A_{i,i+s} = \min\left\{A_{ik} + A_{k+1,i+s} + d_{i-1}d_k d_{i+s}\right\}$$

$$i <= k < i+s \quad \text{for } I = 1,2,\ldots n-s$$

$$A_{02} = \min\begin{cases} A_{00} + A_{12} + d_0 d_1 d_3 & k = 0 \\ A_{01} + A_{22} + d_0 d_2 d_3 & k = 1 \end{cases}$$

$$\min\begin{cases} 0 + 200 + 10\times5\times20 = 1{,}200 \\ 100 + 0 + 10\times2\times20 = \mathbf{500} \end{cases}$$

$$A_{13} = \min\begin{cases} A_{11} + A_{23} + d_1 d_2 d_4 & k = 1 \\ A_{12} + A_{33} + d_1 d_3 d_4 & k = 2 \end{cases}$$

$$\min\begin{cases} 0+480+5\times2\times12=\mathbf{600} \\ 200+0+5\times20\times12=1,400 \end{cases}$$

$$A_{24}=\min\begin{cases} A_{22}+A_{34}+d_2d_3d_5 & k=2 \\ A_{23}+A_{44}+d_2d_4d_5 & k=3 \end{cases}$$

$$\min\begin{cases} 0+960+2\times20\times4=1,120 \\ 480+0+2\times12\times4=\mathbf{576} \end{cases}$$

$$A_{35}=\min\begin{cases} A_{33}+A_{45}+d_3d_4d_6 & k=3 \\ A_{34}+A_{55}+d_3d_5d_6 & k=4 \end{cases}$$

$$\min\begin{cases} 0+2,880+20\times12\times60=17,280 \\ 960+0+20\times4\times60=\mathbf{5,760} \end{cases}$$

For diagonal **s = 3**

$$A_{03}=\min\begin{cases} A_{01}+A_{23}+d_0d_2d_4 & K=1 \\ A_{00}+A_{13}+d_0d_1d_4 & K=0 \\ A_{02}+A_{33}+d_0d_3d_4 & K=2 \end{cases}$$

$$\min\begin{cases} 100+480+10\times2\times12=\mathbf{820} \\ 0+600+10\times5\times12=1,200 \\ 500+0+10\times20\times12=2,900 \end{cases}$$

$$A_{14}=\min\begin{cases} A_{12}+A_{34}+d_1d_3d_5 & K=2 \\ A_{11}+A_{24}+d_1d_2d_5 & K=1 \\ A_{13}+A_{44}+d_1d_4d_5 & K=3 \end{cases}$$

$$\min\begin{cases} 200+960+5\times20\times4=1,560 \\ 0+576+5\times2\times4=\mathbf{616} \\ 600+0+5\times12\times4=840 \end{cases}$$

$$A_{25} = \min \begin{cases} A_{23} + A_{45} + d_2 d_4 d_6 & K=3 \\ A_{22} + A_{35} + d_2 d_3 d_6 & K=2 \\ A_{24} + A_{55} + d_2 d_5 d_6 & K=4 \end{cases}$$

$$\min \begin{cases} 480 + 2{,}880 + 2 \times 12 \times 60 = 4{,}800 \\ 0 + 5{,}760 + 2 \times 20 \times 60 = 8{,}160 \\ 576 + 0 + 2 \times 4 \times 60 = \mathbf{1{,}056} \end{cases}$$

For diagonal **s = 4**

$$A_{04} = \min \begin{cases} A_{00} + A_{14} + d_0 d_1 d_5 & K=0 \\ A_{03} + A_{44} + d_0 d_4 d_5 & K=3 \\ A_{01} + A_{24} + d_0 d_2 d_5 & K=1 \\ A_{02} + A_{34} + d_0 d_3 d_5 & K=2 \end{cases}$$

$$\min \begin{cases} 0 + 616 + 10 \times 5 \times 4 = 816 \\ 820 + 0 + 10 \times 12 \times 4 = 1{,}300 \\ 100 + 576 + 10 \times 2 \times 4 = \mathbf{756} \\ 500 + 960 + 10 \times 20 \times 4 = 2{,}260 \end{cases}$$

$$A_{15} = \min \begin{cases} A_{11} + A_{25} + d_1 d_2 d_6 & K=1 \\ A_{12} + A_{35} + d_1 d_3 d_6 & K=2 \\ A_{13} + A_{45} + d_1 d_4 d_6 & K=3 \\ A_{14} + A_{55} + d_1 d_5 d_6 & K=4 \end{cases}$$

$$\min \begin{cases} 0 + 1{,}056 + 5 \times 2 \times 60 = \mathbf{1{,}656} \\ 200 + 5{,}760 + 5 \times 20 \times 60 = 11{,}960 \\ 600 + 2{,}880 + 5 \times 12 \times 60 = 7{,}080 \\ 616 + 0 + 5 \times 4 \times 60 = 1{,}816 \end{cases}$$

For diagonal **s = 5**

$$A_{05} = \min \begin{cases} A_{00} + A_{15} + d_0 d_1 d_6 & K = 0 \\ A_{01} + A_{25} + d_0 d_2 d_6 & K = 1 \\ A_{02} + A_{35} + d_0 d_3 d_6 & K = 2 \\ A_{03} + A_{45} + d_0 d_4 d_6 & K = 3 \\ A_{04} + A_{55} + d_0 d_5 d_6 & K = 4 \end{cases}$$

$$A_{05} = \min \begin{cases} 0 + 1{,}656 + 10 \times 5 \times 60 = 4{,}656 \\ 100 + 1{,}056 + 10 \times 2 \times 60 = \mathbf{2{,}356} \\ 500 + 5{,}760 + 10 \times 20 \times 60 = 18{,}260 \\ 820 + 2{,}880 + 10 \times 12 \times 60 = 10{,}900 \\ 756 + 0 + 10 \times 4 \times 60 = 3{,}156 \end{cases}$$

i →

j ↓

	0	1	2	3	4	5
5	2356	1656	1056	5760	2880	0
4	756	**616**	576	960	0	
3	820	600	480	0		
2	500	200	0			
1	100	0				
0	0					

S[i, j] – Table

i →

J ↓

	0	1	2	3	4
5	1	1	4	4	4
4	1	1	3	3	
3	1	1	2		
2	1	1			
1	0				

M[i, j] – Table

At each stage of parenthesization, we calculate the minimum cost and is added to obtain the optimal cost.

Thus, the final optimal cost for A_{05} is 2,356 with K value 1.

The parenthesization will be

A0 … 5 = ((A01) (A2 … 5))
= (((A0) (A1)) (A2 … 5))
= (((A0) (A1)) (A2 … 5))
= (((A0) (A1)) ((A2 … 4) (A5)))
= (((A0) (A1)) (((A2 … 3)(A4)) (A5)))
= (((A0) (A1)) (((A2 … 3)(A4)) (A5)))
= (((A0) (A1)) ((((A2) (A3)) (A4)) (A5)))

Example:

First, we have to find the vector P as: P = (2, 3, 7, 5, 6, 4, 3)
where
$P_0=2$
$P_1=3$
$P_2=7$
$P_3=5$
$P_4=6$
$P_5=4$
$P_6=3$

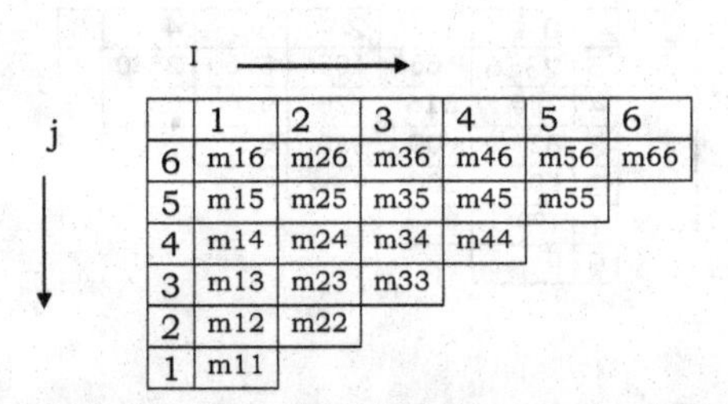

For diagonal **s = 0**

$m_{11}=0$	$m_{22}=0$	$m_{33}=0$	$m_{44}=0$	$m_{55}=0$	$m_{66}=0$.

For diagonal **s = 1**,

$$m_{12} = P_0P_1P_2 = 2*3*7 = 42 \quad K=1$$

$$m_{23} = P_1P_2P_3 = 3*7*5 = 105 \quad K=2$$

$$m_{34} = P_2P_3P_4 = 7*5*6 = 210 \quad K=3$$

$$m_{45} = P_3P_4P_5 = 5*6*4 = 120 \quad K=4$$

$$m_{56} = P_4P_5P_6 = 6*4*3 = 72 \quad K = 5$$

For diagonal **s = 2**,

$$m_{13} = \min\begin{cases} m_{11} + m_{23} + P_0P_1P_3 & k = 1 \\ m_{12} + m_{33} + P_0P_2P_3 & \mathbf{k = 2} \end{cases}$$

$$\min\begin{cases} 0 + 105 + 2\times3\times5 = 135 \\ 42 + 0 + 2\times7\times5 = \mathbf{112} \end{cases}$$

$$m_{24} = \min\begin{cases} m_{22} + m_{34} + P_1P_2P_4 & k = 2 \\ m_{23} + m_{44} + P_1P_3P_4 & \mathbf{k = 3} \end{cases}$$

$$\min\begin{cases} 0 + 210 + 3\times7\times6 = 336 \\ 105 + 0 + 3\times5\times6 = \mathbf{195} \end{cases}$$

$$m_{35} = \min\begin{cases} m_{33} + m_{45} + P_2P_3P_5 & \mathbf{k = 3} \\ m_{34} + m_{55} + P_2P_4P_5 & k = 4 \end{cases}$$

$$\min\begin{cases} 0 + 120 + 7\times5\times4 = \mathbf{260} \\ 210 + 0 + 7\times6\times4 = 378 \end{cases}$$

$$m_{46} = \min\begin{cases} m_{44} + m_{56} + P_3P_4P_6 & \mathbf{k = 4} \\ m_{45} + m_{66} + P_3P_5P_6 & k = 5 \end{cases}$$

$$\min\begin{cases} 0 + 72 + 5\times6\times3 = \mathbf{162} \\ 120 + 0 + 5\times4\times3 = 180 \end{cases}$$

For diagonal **s = 3**,

$$m_{14} = \min\begin{cases} m_{12} + m_{34} + P_0P_2P_4 & K = 2 \\ m_{11} + m_{24} + P_0P_1P_4 & k = 1 \\ m_{13} + m_{44} + P_0P_3P_4 & \mathbf{K = 3} \end{cases}$$

$$\min\begin{cases} 42+210+2\times7\times6=336 \\ 0+195+2\times3\times6=231 \\ 112+0+2\times5\times6=\mathbf{172} \end{cases}$$

$$m_{25}=\min\begin{cases} m_{23}+m_{45}+P_1P_3P_5 & K=3 \\ m_{22}+m_{35}+P_1P_2P_5 & K=2 \\ m_{24}+m_{55}+P_1P_4P_5 & \mathbf{K=4} \end{cases}$$

$$\min\begin{cases} 105+120+3\times5\times4=285 \\ 0+260+3\times7\times4=344 \\ 195+0+3\times6\times4=\mathbf{267} \end{cases}$$

$$m_{36}=\min\begin{cases} m_{34}+m_{56}+P_2P_4P_6 & K=4 \\ m_{33}+m_{46}+P_2P_3P_6 & \mathbf{K=3} \\ m_{35}+m_{66}+P_2P_5P_6 & K=5 \end{cases}$$

$$\min\begin{cases} 210+72+7\times6\times3=408 \\ 0+162+7\times5\times3=\mathbf{267} \\ 260+0+7\times4\times3=344 \end{cases}$$

For diagonal **s = 4**,

$$m_{15}=\min\begin{cases} m_{11}+m_{25}+P_0P_1P_5 & K=1 \\ m_{14}+m_{55}+P_0P_4P_5 & \mathbf{K=4} \\ m_{12}+m_{35}+P_0P_2P_5 & K=2 \\ m_{13}+m_{45}+P_0P_3P_5 & K=3 \end{cases}$$

$$\min\begin{cases} 0+267+2\times3\times4=291 \\ 172+0+2\times6\times4=\mathbf{220} \\ 42+260+2\times7\times4=358 \\ 112+120+2\times5\times4=272 \end{cases}$$

$$m_{26}=\min\begin{cases} m_{22}+m_{36}+P_1P_2P_6 & K=2 \\ m_{23}+m_{46}+P_1P_3P_6 & K=3 \\ m_{24}+m_{56}+P_1P_4P_6 & K=4 \\ m_{25}+m_{66}+P_1P_5P_6 & \mathbf{K=5} \end{cases}$$

$$\min\begin{cases} 0+267+3\times7\times3=330 \\ 105+162+3\times5\times3=312 \\ 195+72+3\times6\times3=321 \\ 267+0+3\times4\times3=\mathbf{303} \end{cases}$$

For diagonal **s = 5**,

$$m_{16}=\min\begin{cases} m_{11}+m_{26}+P_0P_1P_6 & K=1 \\ m_{12}+m_{36}+P_0P_2P_6 & K=2 \\ m_{13}+m_{46}+P_0P_3P_6 & K=3 \\ m_{14}+m_{56}+P_0P_4P_6 & K=4 \\ m_{15}+m_{66}+P_0P_5P_6 & \mathbf{K=5} \end{cases}$$

$$m_{16}=\min\begin{cases} 0+303+2\times3\times3\ =321 \\ 42+267+2\times7\times3=351 \\ 112+162+2\times5\times3=304 \\ 172+72+2\times6\times3=280 \\ 220+0+2\times4\times3=\mathbf{244} \end{cases}$$

	1	2	3	4	5	6
6	244	303	267	162	72	0
5	220	**267**	260	120	0	
4	172	195	210	0		
3	112	105	0			
2	42	0				
1	0					

S[i, j] – Table: The entries in the S –Table are being filled by observing the K value of Min(M[i, j]) values.

	1	2	3	4	5
6	5	5	3	4	5
5	4	4	3	4	
4	3	3	3		
3	2	2			
2	1				

M[i, j] – Table
At each stage of parenthesization, we calculate the minimum cost and is added to obtain the optimal cost. Thus, the final optimal cost for M_{16} is 244.

Suppose we want to multiply all the matrices i.e., $M_{1 \ldots 6}$
Therefore, for that, we will have to use the S table to choose the way to multiply.

For M_{16} I = 1, J = 6
Since i! = j so
X = Optimal Parenthesis (A, S, 1, 5) and Y = Optimal Parenthesis (A, S, 6, 6)
For Y = Optimal Parenthesis (A, S, 6, 6) sin i = j so no operation will perform.

Again for X= Optimal Parenthesis (A, S, 1, 5) i = 1 and j = 5 so
X = Optimal Parenthesis (A, S, 1, 4) and Y = Optimal Parenthesis (A, S, 5, 5)
For Y = Optimal Parenthesis (A, S, 5, 5) sin i = j so no operation will perform.

Again for X= Optimal Parenthesis (A, S, 1, 4) I =1 and j = 4 so X = Optimal Parenthesis (A, S, 1, 3) and Y = Optimal Parenthesis (A, S, 4, 4)
For Y = Optimal Parenthesis (A, S, 4, 4) sin i = j so no operation will perform.

Again for X= Optimal Parenthesis (A, S, 1, 3) i = 1 and j = 3 so X = Optimal Parenthesis (A, S, 1, 2) and Y = Optimal Parenthesis (A, S, 3, 3)
For Y = Optimal Parenthesis (A, S, 3, 3) sin i = j so no operation will perform.

Again for X= Optimal Parenthesis (A, S, 1, 2) i = 1 and j = 2 so X = Optimal Parenthesis (A, S, 1, 1) and Y = Optimal Parenthesis (A, S, 2, 2)
For Y = Optimal Parenthesis (A, S, 2, 2) sin i = j so no operation will perform.

Since for X = Optimal Parenthesis (A, S, 1, 1) i = 1 and j = 1 and Y = Optimal Parenthesis (A, S, 2, 2) i = 2 and j = 2 so all the functions will be called recursively by using stack.

Therefore, finally to multiply $M_1 \ldots {}_6$ the total number of steps required is 244 and the process will be

$$
\begin{aligned}
&m_{15} + m_{66} + P_0P_5P_6 \\
&= \left(m_{14} + m_{55} + P_0P_4P_5\right) + m_{66} + P_0P_5P_6 \\
&= \left(\left(\left(m_{13} + m_{44} + P_0P_3P_4\right) + m_{55} + P_0P_4P_5\right)\right) + m_{66} + P_0P_5P_6 \\
&= \left(\left(\left(\left(m_{12} + m_{33} + P_0P_2P_3\right) + m_{44} + P_0P_3P_4\right) + m_{55} + P_0P_4P_5\right)\right) + m_{66} + P_0P_5P_6
\end{aligned}
$$

4.3.1.2 Elements of Dynamic Programming

1. Optimal substructure
2. Overlapping subproblem
3. Memorization.

4.4 LONGEST COMMON SUBSEQUENCE PROBLEM

A subsequence is any subset of the elements of a sequence that maintains the same relative order. If A is a subsequence of B, then this is denoted as $A \subset B$.

1. **Characterizing the longest common subsequence.**
   Optimal Substructure of an LCS
   Let X=<X1, X2, ..., Xm> and Y=<Y1, Y2, ... Yn> be sequences and let Z=[Z1,Z2,...Zk] be any LCS of X and Y

   1. If Xm=Yn then Zk=Xm=Yn and Zk−1 is an LCS of Xm−1 and Yn−1
   2. If Xm ≠ Yn then Zk ≠ Xm implies that Z is an LCS of Xm−1 and Y.
   3. If Xm ≠ Yn then Zk ≠ Yn implies that Z is an LCS of X and Yn−1

2. **Recursive Solution**

   The recursive solution to the LCS problem involves establishing a recurrence for the value of an optimal solution.

   Let us define c[i, j] to be the length of an LCS of the sequence Xi and Yj. If either i=0 or J=0, one of the sequences has length 0, so the LCS has length 0. The optimal substructure of the LCS problem gives the formula as

$$C[i,j] = \begin{cases} 0 & \text{if } i=0 \text{ and } j=0 \\ c[i-1,j-1]+1 & \text{if } i,j>0 \text{ and } Xi=Yj \\ \text{Max}(c[i,j-1],c[i-1,j]) & \text{if } i,j>0 \text{ and } Xi \neq Yj \end{cases}$$

3. **Computing the Length of an LCS**

   Here two sequences X={X1, X2 ... Xm} and Y={Y1, Y2, ... Yn} are passed as input to the procedure. It stores the c[i, j] values in a table C[0 ... m, 0 ... n] whose entries are computed in a row-major order. It also maintains the b[1 ... m, 1 ... n] to simplify the construction of an optimal solution.

ALGORITHM LCS-Length(X,Y)

```
 1. m ← length[X]
 2. n ← length[Y]
 3. for i ← 1 to m
 4. do c[i,0] ← 0
 5. for j ← 1 to n
 6. do c[0,j] ← 0
 7. for i ← 1 to m
 8. for j ← 1 to n
 9. do if (Xi = Yj)
10. then c[i,j] ← c[i-1,j-1] + 1
11. b[i,j] ← "↖"
12. else if c[i-1,j] ≥ c[i,j-1]
13. then c[i,j] ← c[i-1,j]
14. b[i,j] ←"↑"
15. else
16. c[i,j] ← c[i,j-1]
17. b[i,j] ← "←"
18. return c and b
```

4. Constructing an LCS

The b table returned by the above algorithm can be used to quickly construct an LCS of X={X1, X2 … Xm} and Y={Y1, Y2, … Yn}. We simply begin at b[m, n]and trace through the table according to the arrowheads.

Whenever we encounter a ↖ in the entry b[i, j]. It implies that Xi=Yi is an element of LCS. The element of the LCS is encountered in reverse order by this method and for other arrows move according to their direction.

ALGORITHM Print-LCS(b,X,i,j)

```
1. if i=0 or j=0
2. then return
3. if b[i,j] = "↖".
4. then Print-LCS(b,X,i-1,j-1)
5. print Xi
6. else if b[i,j] = "↑"
7. then Print-LCS(b,X,i-1,j)
8. else Print-LCS(b,X,i,j-1)
```

Example:

Let X={C, O, W}
And Y={B, R, O, W, N}

		0	1	2	3	4	5
		Yj	B	R	**O**	**W**	N
0	Xi	0	0	0	0	0	0
1	C	0	0	0	0	0	0
2	**O**	0	0	0	1↖	1←	1←
3	**W**	0	0	0	1↑	2↖	2←

Example:

Let X={PROVIDENCE} and Y={PRESIDENT}

		0	1	2	3	4	5	6	7	8	9	10
		Yj	P	R	O	V	I	D	E	N	C	E
0	Xi	0	0	0	0	0	0	0	0	0	0	0
1	P	0	1↖	1←	1←	1←	1←	1←	1←	1←	1←	1←
2	R	0	1↑	2↖	2←	2←	2←	2←	2←	2←	2←	2←
3	E	0	1↑	2↑	2↑	2↑	2↑	2↑	3	3←	3←	3
4	S	0	1↑	2↑	2↑	2↑	2↑	2↑	3↑	3↑	3↑	3↑
5	I	0	1↑	2↑	2↑	2↑	3↖	3↑	3↑	3↑	3↑	3↑
6	D	0	1↑	2↑	2↑	2↑	3↑	4↖	4←	4←	4←	4←
7	E	0	1↑	2↑	2↑	2↑	3↑	4↑	5↖	5←	5←	5
8	N	0	1↑	2↑	2↑	2↑	3↑	4↑	5↑	6↖	6←	6←
9	T	0	1↑	2↑	2↑	2↑	3↑	4↑	5↑	6↑	6↑	6↑

Example:

Let X={CLASSIFIER} and Y={CLASSIFIED}

		0	1	2	3	4	5	6	7	8	9	10
		Yj	C	L	A	S	S	I	F	I	E	D
0	Xi	0	0	0	0	0	0	0	0	0	0	0
1	C	0	1↖	1←	1←	1←	1←	1←	1←	1←	1←	1←
2	L	0	1↑	2↖	2←	2←	2←	2←	2←	2←	2←	2←
3	A	0	1↑	2↑	3↖	3←	3←	3←	3←	3←	3←	3←
4	S	0	1↑	2↑	3↑	4↖	4↖	4←	4←	4←	4←	4←
5	S	0	1↑	2↑	3↑	4↖	5↖	5←	5←	5←	5←	5←
6	I	0	1↑	2↑	3↑	4↑	5↑	6↖	6←	6↖	6←	6←
7	F	0	1↑	2↑	3↑	4↑	5↑	6↑	7↖	7←	7←	7←
8	I	0	1↑	2↑	3↑	4↑	5↑	6↖	7↑	8↖	8←	8←
9	E	0	1↑	2↑	3↑	4↑	5↑	6↑	7↑	8↑	9↖	9←
10	R		1↑	2↑	3↑	4↑	5↑	6↑	7↑	8↑	9↑	9←

4.5 DIVIDE AND CONQUER VS. DYNAMIC PROGRAMMING

- Both techniques split their input into parts, find subsolutions to the parts, and synthesize larger solutions from smaller ones.
- Divide-and-conquer splits its input at prespecified deterministic points (e.g., always in the middle).
- Dynamic programming splits its input at every possible split point rather than at a pre-specified point. After trying all split points, it determines which split point is optimal.

4.6 QUESTIONS

4.6.1 Short Questions

1. What are top-down and bottom-up in dynamic programming?
2. Write the elements of dynamic programming.
3. What is LCS?
4. What is dynamic programming?
5. What is bottom-up computation?
6. Define optimal substructure.
7. Define matrix chain multiplication.
8. Explain elements of dynamic programming.
9. How to characterize the LCS?
10. Differentiate between divide and conquer versus dynamic programming.

4.6.2 Long Questions

1.

M_1	M_2	M_3	M_4	M_5
15×10	10×25	25×15	15×5	5×30

Find the matrix chain multiplication $M_{1 \ldots 5}$.

2. Write an algorithm for LCS and give an example.
3. Discuss dynamic programming.
4. Discuss how to develop dynamic programming algorithms.
5. Write the principle of optimality.

1. Construct and LCS let X = {CLASSIFIER} and Y = {CLASSIFIED}.
2. Discuss matrix chain multiplication and how to order the multiplications.
3. Write algorithm for matrix multiplication.
4. Write algorithm for matrix chain order.
5. Short note on matrix chain multiplication and matrix multiplication.

CHAPTER 5

Greedy Algorithms

5.1 GREEDY DESIGN APPROACH

This is another type of method for designing the algorithm. Like dynamic programming, each greedy algorithm consists of two parts:

- **Optimal substructure:**
  An optimal solution to the problem contains an optimal solution to subproblems.
- **Greedy choice property:**
  A global optimum can be arrived at by selecting a local optimum.
  The first property may make greedy algorithms look like dynamic programming. However, the two techniques are quite different.

5.1.1 Characteristics and Features of Problems Solved by Greedy Algorithms

To construct the solution optimally, the algorithm maintains two sets as

- **SET1:** Contains Chosen Items
- **SET2:** Contains Rejected Items

DOI: 10.1201/9781003093886-5

A greedy algorithm consists of four functions to find the optimal solution:

1. A function that checks whether the chosen set of items provides a solution.
2. A function that checks the feasibility of a set.
3. The selection function tells which of the candidates is the most promising.
4. An objective function that does not appear explicitly gives the value of a solution.

5.1.2 Basic Structure of Greedy Algorithm

1. Initially, the set of chosen items is empty, i.e., solution set.
2. At each step
    - Item will be added in a solution set by using function.
    - If the set would no longer be feasible
        i. Reject items under consideration (and never consider again)
    - Else if a set is still feasible, then
        - Add the current item.

5.1.3 What Is Feasibility

A feasible set is promising if it can be extended to produce not merely a solution, but an optimal solution to the problem.

The greedy strategy usually progresses in a top-down fashion, making one greedy choice after another, reducing each problem to a smaller one.

5.1.3.1 How to Prove Greedy Algorithms Optimal

An optimization problem is one in which you want to find, not just *a* solution, but the *best* solution

A "greedy algorithm" sometimes works well for optimization problems

A greedy algorithm works in phases. At each phase:

- You take the best you can get right now, without regard for future consequences

- You hope that by choosing a *local* optimum at each step, you will end up at a *global* optimum

For proving "greedy algorithm" to be optimal, we need to prove

- Optimal substructure
- Greedy choice property.

The procedure is as follows:

a. Consider a globally optimal solution.
b. Show greedy choice at first step reduces the problem to the same but a smaller problem.
   Greedy choice must be:
   i. Part of an optimal solution and
   ii. Can be made first.
c. Use induction to show greedy choice is best at each step, i.e., optimal substructure.

> An approach is called greedy when we make decisions for each step based on what seems best at the current step.

5.2 AN ACTIVITY – SELECTION PROBLEM

In activity scheduling problem, we have to determine the optimum number of activities that are scheduled to the resources.

Given a set $S = \langle 1, 2, 3, \ldots, n \rangle$ of n number of activities each is scheduled to use some resource. Where each condition might arise where multiple activities have their start and finish time as denoted as Si and Fi.

A condition might arise where multiple activities are scheduled to have a common resource or the start or finish time of activities may overlap.

From the given set of activities, we can say that Ai and Aj are noninterfering activities, if and only if they both have different start and finish time, that means their start and finished do not overlap Ai(Si, Fi) and Aj(Sj, Fj) so for noninterfering

$$Ai \cap Aj = \phi$$

The main aim of the A-S problem is to select a maximum possible set of activities where each activity is non-interfering.

For optimal greedy, first select the activity with minimum duration (Fi – Si) and scheduled it.

Next, we skip all the activities that interfere with this one that means we have to select noninterfering activity having a minimum duration and then we have to schedule it. This process will repeat until all the activities are considered.

It will be faster if we assume that the activity is arranged in increasing order of their finishing time, i.e., $f_1 \leq f_2 \leq f_3 \leq \ldots \leq f_n$

Algorithm Activity-Schedule(S,F)

```
1.    Set AList ← 1 (Activity 1 is scheduled first)
2.    Pre ← 1
3.    Loop & Checking interference and appending list
4.    for I ← 2 to n
5.    if (S[i] >= f[pre]) then
6.    {
7.    Append i to Alist (Scheduled i next)
8.    Set Pre ← i
9.    }
10.   Return Alist
```

In this algorithm, the activities are first sorted according to their finishing time, from the earliest to the latest, where a tie can be broken arbitrarily. Then, the activities are greedily selected by going down the list and by picking whatever activity that is compatible with the current selection.

The sorting part can be as small as O(n log n) and the other part is O(n), so the total is O(n log n).

Finally, the activity selection problem is the problem of selecting the largest set of mutually compatible activities.

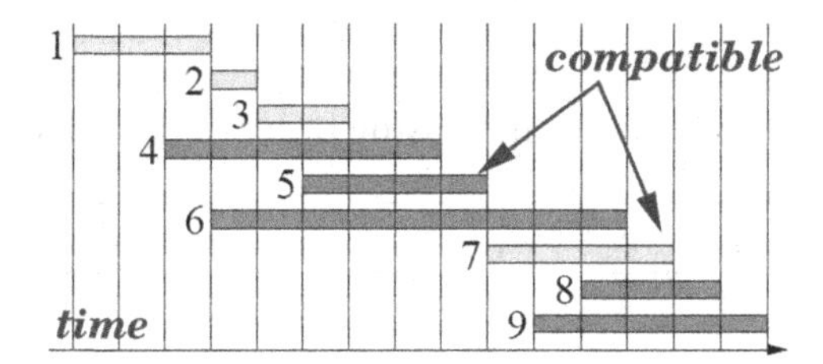

Example:

Consider the following six activities:

start[] = {1, 3, 0, 5, 8, 5};

finish[] = {2, 4, 6, 7, 9, 9};

The maximum set of activities that can be executed by a single person is {0, 1, 3, 4}.

The greedy choice is to always pick the next activity whose finish time is least among the remaining activities and the start time is more than or equal to the finish time of the previously selected activity. We can sort the activities according to their finishing time so that we always consider the next activity as the minimum finishing time activity.

1. Sort the activities according to their finishing time.
2. Select the first activity from the sorted array and print it.
3. Do the following for the remaining activities in the sorted array.
   a. If the start time of this activity is greater than the finish time of the previously selected activity, then select this activity and print it.

Example:

Given ten activities along with their start and finish time as

S = $<A_1, A_2, A_3, A_4, A_5, A_6, A_7, A_8, A_9, A_{10}>$

Si = <1, 2, 3, 4, 5, 6, 7, 8, 9, 10>

F = <5, 3, 4, 6, 7, 8, 11, 10, 12, 13>.

Solution:

Arrange the activities in increasing order of the finish time as

Activity	A_2	A_3	A_1	A_4	A_5	A_6	A_8	A_7	A_9	A_{10}
Start	2	3	1	4	5	6	8	7	9	10
Finish	3	4	5	6	7	8	10	11	12	13

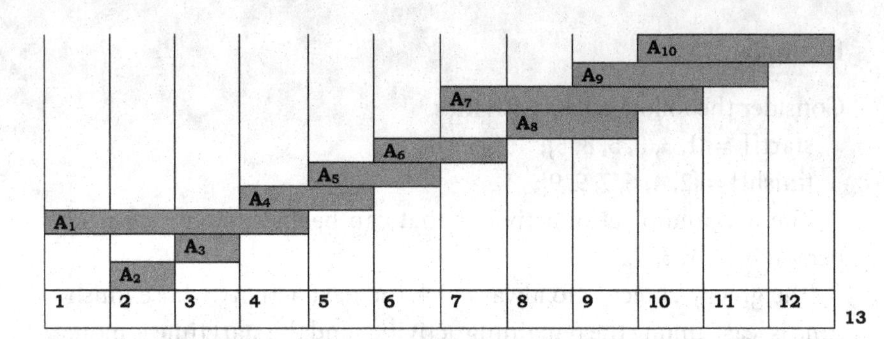

- Select A2 next A3 as both are noninterfering and add them to the activity list.
- Since A1 interfere so reject it.
- Next Select A4 and add it to activity list.
- Since A5 interfere so reject it.
- Next, select A6 and A8 and add them to the activity list.
- Since A7 and A9 interfere so again reject it.
- Next, select A10 and add it to activity list.

The solution is A_2, A_3, A_4, A_6, A_8 and A_{10}.

Example:

Consider 11 activities sorted by finish time:

(1, 4), (3, 5), (0, 6), (5, 7), (3, 8), (5, 9), (6, 10), (8, 11), (8, 12), (2, 13), (12, 14)

Arrange the activities in increasing order of the finish time as

Activity	A_1	A_2	A_3	A_4	A_5	A_6	A_7	A_8	A_9	A_{10}	A_{11}
Start	1	3	0	5	3	5	6	8	8	2	12
Finish	4	5	6	7	8	9	10	11	12	13	14

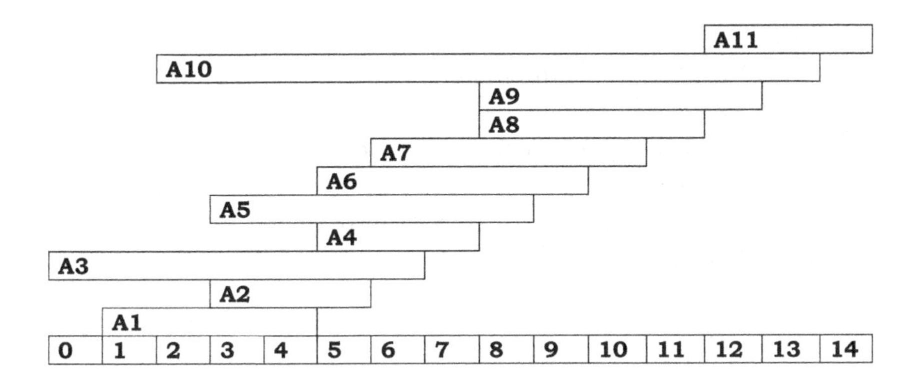

By applying the algorithm, the solution is $\mathbf{A_1}$, $\mathbf{A_4}$, $\mathbf{A_8}$ and $\mathbf{A_{11}}$.

Recursive Algorithm For Activity Selection Problem
Recursive_activity_selector(S, F, K, N)

```
1. m := k+1
2. while m<=n and s[m] < f[k]  #find the first
   activity in Sk to finish
3.    m := m+1
4. If m<=n
5. RETURN {Am} U RECURSIVE _ACTIVITY_SELECTOR(s,f,m,n)
6. Else  return NULL
```

Example:

Consider the set of actives Ai with their start times Si and finish times fi

I	1	2	3	4	5	6	7	8	9
Si	1	4	5	7	8	10	7	10	12
Fi	5	6	7	8	9	11	12	14	16

According to algorithm,
S: Starting time
F: Finish time
K: Initial Activity i.e./1
N: 9 (Number of activities)

So m = k + 1 = 2
While(m<=n and S[m] <F[k])

M = m+a
So here (2<9 and S[2] <F[1]) throughout/4<5 (True) i.e/m = m+1 = 3

Now (3<=9 and S[3] <F[1]) i.e/5<5 (False) so loop will terminate
Now if(m<=n) i.e/ 3<=9 true
So return {A3} U Recursive_Activity_Selector(s,f,m,n)
Now for this call K = 3
M = k+1 = 4

While(m<=n and S[m] < F[k]) i.e/ 4<=9 and 7< 7 (False)
So loop will terminate.

Now if(m<=n) i.e/ 4<=9 True
So return {A4} U Recursive_Activity_Selector(s,f,m,n)

So for this call K=4
M = k+1 = 5
While(m<=n and S[m] < F[k]) i.e/ 5<=9 and 8<8 (False)
So loop will terminate.
Now if(m<=n) i.e/ 5<=9 True
So return {A5} U Recursive_Activity_Selector(s,f,m,n)

For this call K=5
M = k+1 = 6
While(m<=n and S[m] < F[k]) I.E/ 6<=9 AND 10<9 (False)
So loop will terminate.
Now if (6<=9) True
So return {A6} U Recursive_Activity_Selector(s,f,m,n)

For this call k =6
M = k +1 = 7
While(m<=n and S[m] < F[k]) I.E/ 7<=9 AND 7<11 (True) so m = m+1= 8

While(m<=n and S[m] < F[k]) I.E/ 8<=9 AND 10<11 (True) so m = m+1= 9

While(m<=n and S[m] < F[k]) I.E/ 9<=9 AND 12<11 (False)
so loop will terminate.

W if m<=n i.e/ 9<= 9 True
So return {A9} U Recursive_Activity_Selector(s,f,m,n)

For this call k =9
M=k+1 = 10

While(m<=n and S[m] < F[k]) False
So loop will terminate
Now if (m<=n) i.e/10<=9 False
Therefore, the procedure will stop the execution.
Since it is a recursive call, it will return as LIFO concept (STACK)
Now the final activities are A1, A3, A4, A5, A6 and A9.

5.3 KNAPSACK PROBLEM

The **knapsack problem** or **rucksack problem** is a problem in combinatorial optimization: Given a set of items, each with a mass and a value, determine the number of each item to include in a collection so that the total weight is less than or equal to a given limit and the total value is as large as possible. It derives its name from the problem faced by someone who is constrained by a fixed-size knapsack and must fill it with the most valuable items.

There are two types of knapsack problems:

1. **0/1 Knapsack Problem**: Here the items may not be broken into smaller pieces, so thief may decide either to take an item or to leave it (binary choice), but may not take a fraction of an item. 0/1 Knapsack Problem does not exhibit greedy algorithm it only follows a dynamic programming algorithm.

2. **Fractional Knapsack Problem**: Here the thief can take fractions of items, meaning that the items can be broken into smaller pieces so that the thief may decide to carry only a fraction of Xi of item I, where $0 \leq Xi \leq 1$. Fractional Knapsack Problem can be solved by "Greedy Choice Property".

The knapsack problem can be solved in three different criterions:

- Greedy Criterion – I

  In this, the items are arranged by their values. Here the item with maximum value is selected first and the process continues till the minimum value.

- Greedy Criterion – II

  In this, the items are arranged by their weights. Here the item with the largest weight is selected first and the process continues till the maximum weight.
- Greedy Criterion – III

  In this, the item is arranged by the certain ratio Pi, where Pi is the ratio of values over weight. Here selection process from maximum ratio to minimum ratio.

After taking the greedy criterion, we check whether the weight Wi of items Ii from the top of the list is less than the Knapsack Capacity W or not, if Wi is less than or equal to W then the item is selected and the whole process continues for $I_{(i+1)}$, $I_{(i+2)}$, … along the maximum weight W – Wi otherwise the item having Wi is ignored and the process continues with $I_{(i+1)}$, $I_{(i+2)}$,… along with weights.

Algorithm Knapsack(I, W, V)

```
1. Set S←{ϕ}
2. Set V ← 0
3. for I ← 1 to N [loop, adding items to set S]
4. if (Wi ≤ W) then
5. Set S ← Ii (Adding items to Sets)
6. Set V ← V + Vi [Adding the value]
7. Set W ← W - Wi [Net maximum weight]
8. Return S and V
```

Example:

$I = \{I_1, I_2, I_3\}$

$W = \{5, 4, 3\}$

$V = \{6, 5, 4\}$

Weight of knapsack is W = 7

Based on criterion – 1

Initially

Item	W	V
I_1	5	6
I_2	4	5
I_3	3	4

Item I_1 is selected first as it has the maximum value. Next, all the other items result in overflow, so by this criterion, we get the maximum value as 6 having a total weight of 5 which is less than W.

So now based on Criterion – II

Initially

Item	W	V
I_1	5	6
I_2	4	5
I_3	3	4

Arrange the weights according to weight (ascending order)

Item	W	V
I_3	3	4
I_2	4	5
I_1	5	6

First item with least weight is selected, i.e., I_1.

Now the new weight of Knapsack is $W - W_i = 7 - 3 = 4$.

Next item with weight $W_i = 4$ is selected as the total weight is equal to the knapsack capacity W; thus, the maximum value we get is $V = 9$.

Fractional Knapsack

Algorithm Fractional_knapsack(W, V, W)

```
1.  for i ← 1 to n
2.  do x[i] ← 0
3.  Weight ← 0
4.  While weight ← W
5.  do i ← best remaining item
6.  if weight + w[i] ≤ W
7.  then x[i] ← 1
8.  weight ← weight + w[i]
9.  else
10. x[i] ←(w - weight)/ w[i]
11. weight ← W
12. return x
```

Consider five items along with their respective weights and values:

$I = \{I_1, I_2, I_3, I_4, I_5\}$

$W = \{5, 10, 20, 30, 40\}$

$V = \{30, 20, 100, 90, 160\}$

The knapsack has capacity W = 60.

Item	W	V
I_1	5	30
I_2	10	20
I_3	20	100
I_4	30	90
I_5	40	160

Item	W	V	$P_i = (V_i/W_i)$
I_1	5	30	6.0
I_2	10	20	2.0
I_3	20	100	5.0
I_4	30	90	3.0
I_5	40	160	4.0

Arrange the above items according to the descending order with P_i

Item	W	V	$P_i = (V_i/W_i)$
I_1	5	30	6.0
I_2	20	100	5.0
I_3	40	160	4.0
I_4	30	90	3.0
I_5	10	20	4.0

Now filling the knapsack by the decreasing value of P_i

Maximum value is $V_1 + V_2 + \text{New}(V_3)$:

$= 30 + 100 + (35 * 4.0)$

$= 270.$

5.4 HUFFMAN ENCODING

This is used for the compression of the data. Huffman code is used to encode or decode the data. In general, it's too difficult to pass the data in a low-bandwidth transmission. Therefore, it will be better to encode the data into a binary format so that it will be easier to transmit.

Example:

Let	A	B	C	D	E	F
Frequency (in thousand)	45	13	12	16	9	5
Fixed-length code word	000	001	010	011	100	101
Variable-length code word	0	101	100	111	1101	1100

If we use fixed length code, then we need 3 bits to represent 6 characters:

i.e., A – 000, B – 001, C – 010, D – 011, E – 100, F – 101

Therefore, this method requires 300,000 bits to code the entire file having 100,000 characters.

If we use variable length code, we can do considerably better than the fixed length code by giving frequent characters short code words and infrequent characters long code word.

From the above, this code requires

$(45 * 1 + 13 * 3 + 12 * 3 + 16 * 3 + 9 * 4 + 5 * 4) * 1{,}000 = 224{,}000$

bits to represent a file, i.e., it saves approximately saves 25%.

Prefix Codes: This code can also be called prefix-free code. The standard solution for unique decoding is to insist that the code word be prefix free.

This means that if a, b $\in \Sigma$, a $\neq$ b then code word for a is not a prefix of the code word for b and vice versa.

Encoding is always simple for any binary character code. We just concatenate the code words representing each character of the file.

For example, we code the three-character file abc as 0.101.100 = 0101100, whereas we use (.) for concatenation.

Prefix codes are quite preferable because they simplify the decoding.

Since no code word is a prefix of any other, the code word that begins an encoded file is unambiguous.

Variable length code representation having more benefits than fixed length code.

Example:

Fixed-Length Code Representation

	a	b	c	d	e	f	g
Frequency	37	18	29	13	30	17	6
Fixed-length code word	000	001	010	011	100	101	110

- Total size is:
  $(37 + 18 + 29 + 13 + 30 + 17 + 6) \times 3 = 450$ bits.

Variable-Length Code Representation

	a	b	c	d	e	f	g
Frequency	37	18	29	13	30	17	6
Variable-length code	10	011	111	1101	00	010	1100

- Total size is:
  $37 \times 2 + 18 \times 3 + 29 \times 3 + 13 \times 4 + 30 \times 2 + 17 \times 3 + 6 \times 4 = 402$ bits.
- A savings of approximately 11%

Encoding: For encoding, concatenate the code words representing each character of the file as

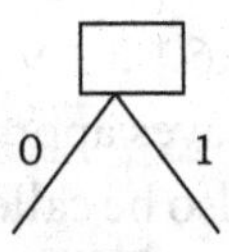

i.e., left represents 0 and right as 1.

Decoding: For decoding, remove it from the encode file and repeatedly parse, i.e., when

0 is found then go to left.

0 is found then go to right.

5.4.1 Prefix-Free Code Representation

Binary tree representation of prefix-free binary code:

- 0 = label (left branch) and 1 = label (right branch).

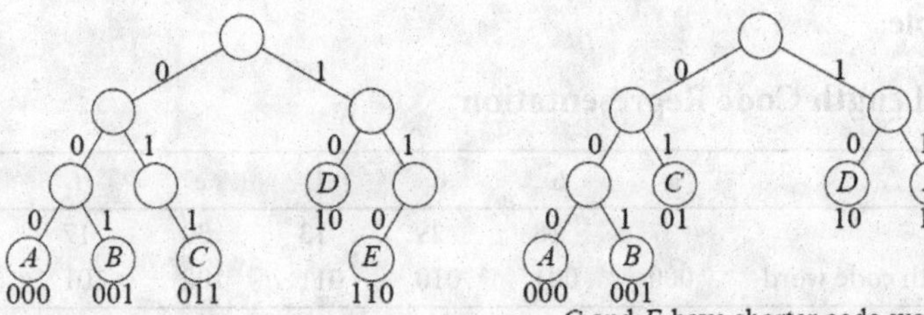

C and *E* have shorter code-word;
each non-terminal node has 2 children.

Algorithm Huffman (C)

```
1.   n ← |C|
2.   Q ← C    [Q is a binary Heap]
3.   for i ← 1 to n-1     [O(n) build Heap]
4.   Z ← allocate node() [O(n)]
5.   X ← Extract_Min(Q)   [O(lgn)O(n) times]
6.   Y ← Extract_Min(Q)   [O(lgn)O(n) times]
7.   Left[Z] ← X
8.   Right[Z] ← Y
9.   f(Z) ← f(X) + f(Y)
10.  Insert (Q,Z)         [O(lgn)O(n) times]
11.  ReturnExtract_Min(Q)   ----------------------------
                           O(n lgn)
```

Example:

$\sum_{0}\{A, B, \ldots, E\}$ and $p(A) = 0.1 = p(B)$, $p(C) = 0.3$, $p(D) = p(E) = 0.25$.

The nodes in S are shown shaded.

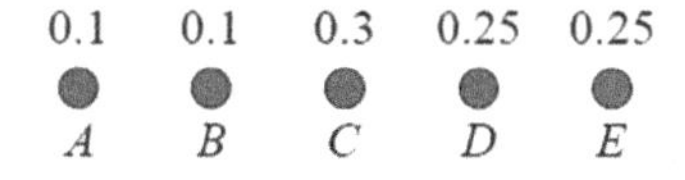

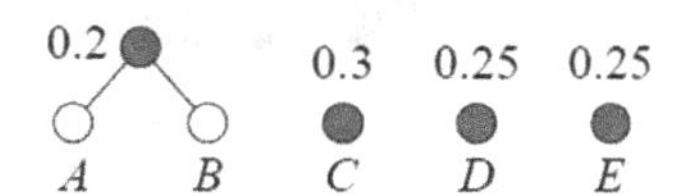

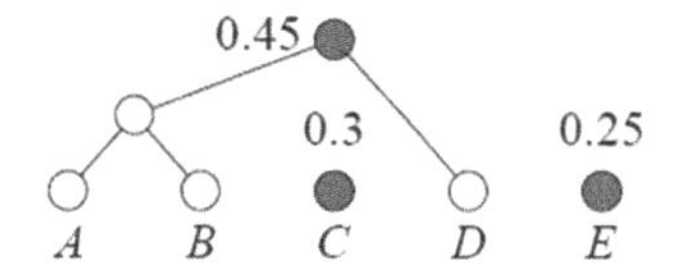

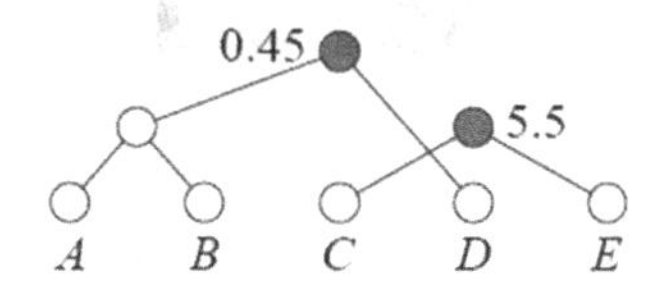

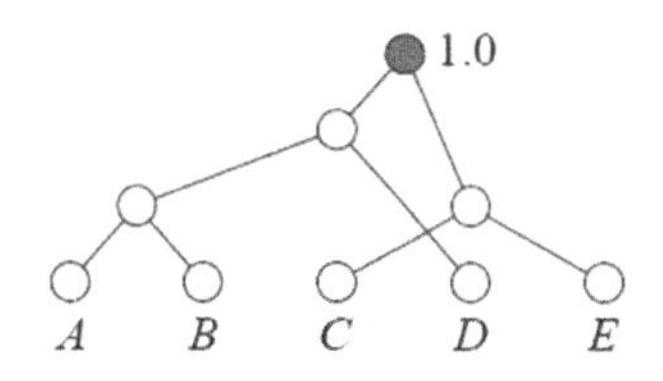

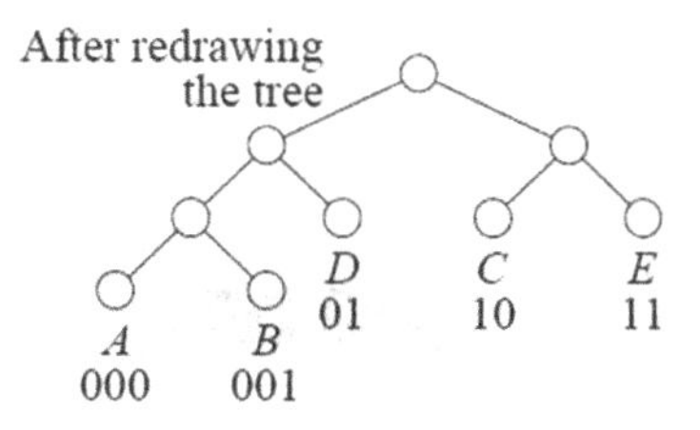

Example:

Find an Optimal Huffman Code for the following set of frequencies:

A = 50	B = 25	C = 15	D = 40	E = 75

Solution:

Given C = {A, B, C, D, E}
F(C) = {50, 25, 15, 40, 75}
n = 5
Q = C

A = 50	B = 25	C = 15	D = 40	E = 75

For i = 1 to n – 1 i.e./I = 1 to 4
Z = allocate a memory X = Extract_Min(Q) = C|15
Y = Extract_Min(Q) = B|25
Left [Z] = X Right[Z] = Y f(Z) ← f(X) + f(Y)

Example:

Construct the Huffman tree

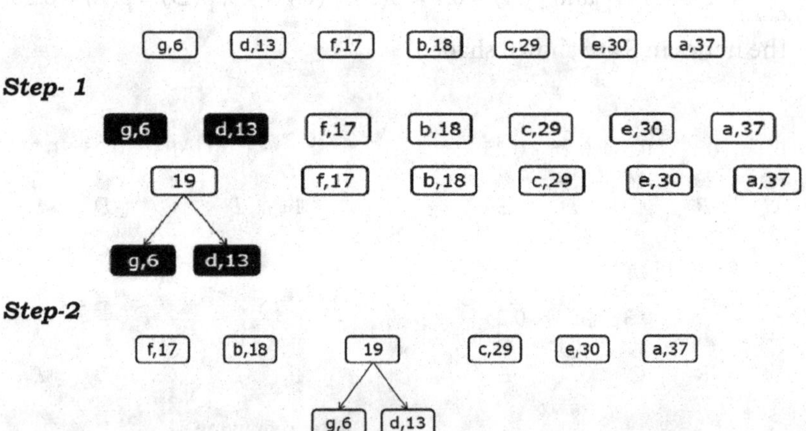

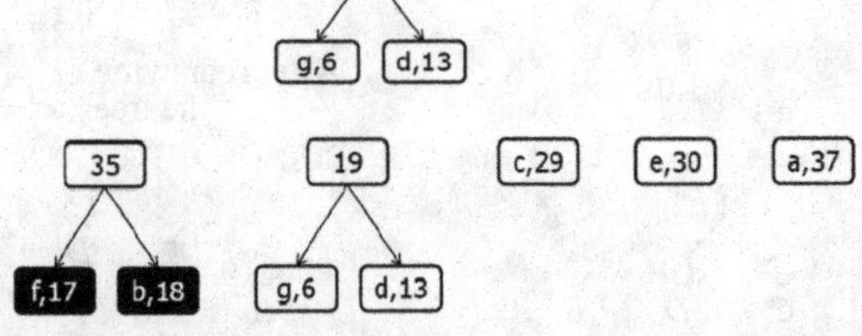

Step- 3

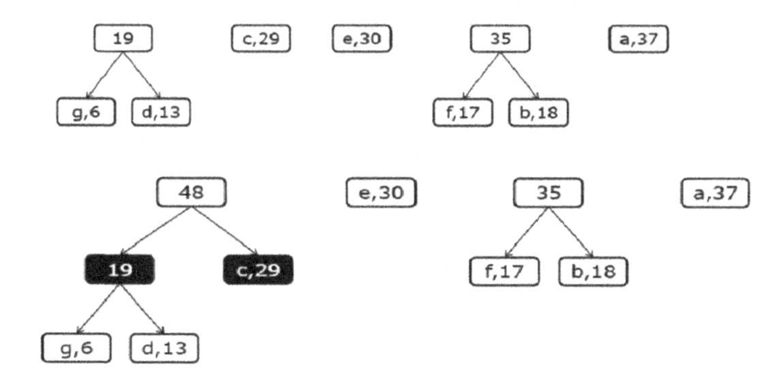

Step - 4

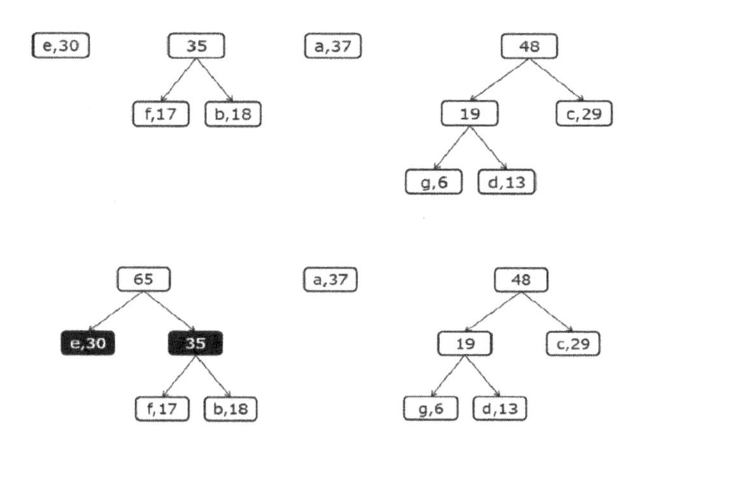

Step-5

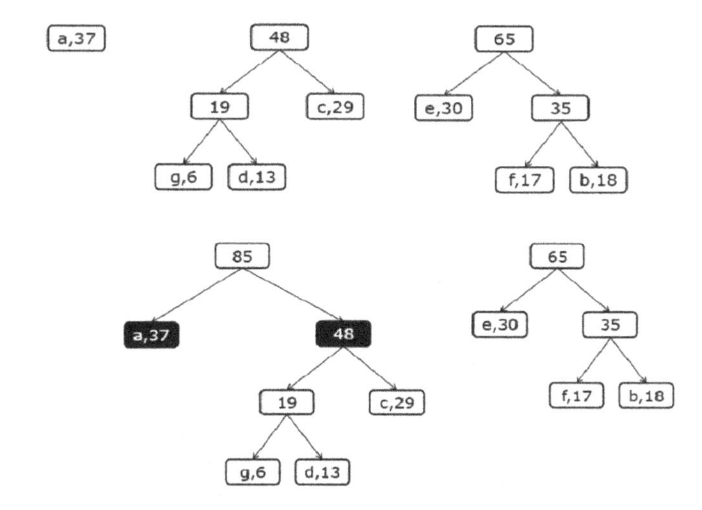

Step - 6

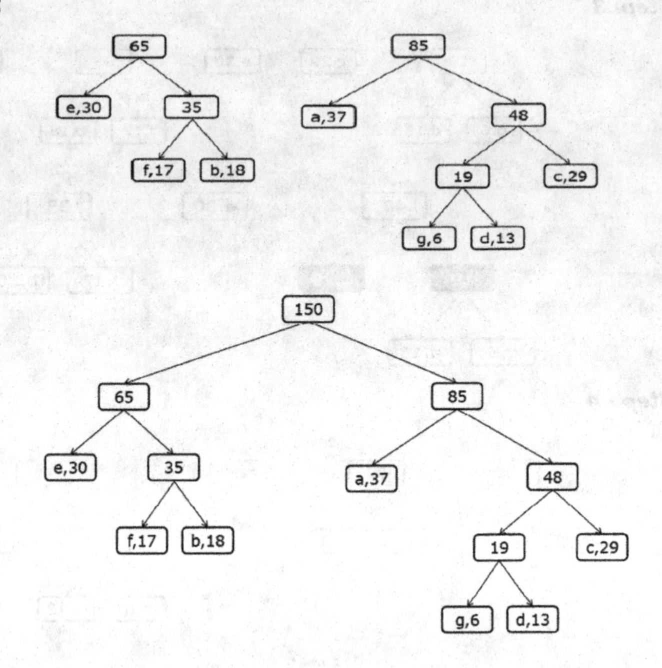

Resulting Codes

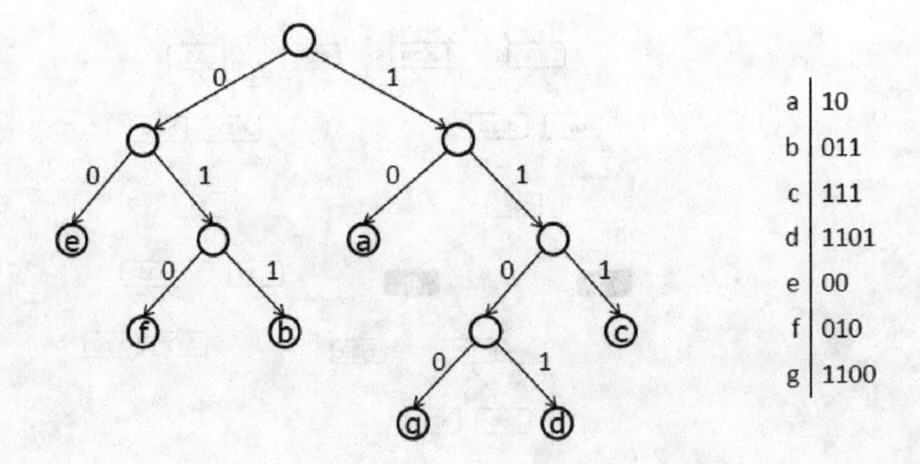

Example:

Huffman tree is used for the compression of the data. Huffman code is used to encode or decode the data. In general, it's too difficult to pass the data in a low-bandwidth transmission. Therefore, it will be better to encode the data into a binary format so that it will be easier to transmit.

Given the string as

"streets are stone stars are not"

Count the frequencies of individual characters.

Character	s	t	r	e	a	o	n
Frequency	5	5	4	5	3	2	2

Solution:

Given C = {s, t, r, e, a, o, n}

F(C) = {5, 5, 4, 5, 3, 2, 2}

n = 7

Q = C

For i = 1 to n − 1 i.e/I = 1 to 6

Z = allocate a memory   X = Extract_Min(Q) = n|2

Y = Extract_Min(Q) = o|2

Left [Z] = X   Right[Z] = Y   f(Z) ← f(X) + f(Y)

So the tree will be

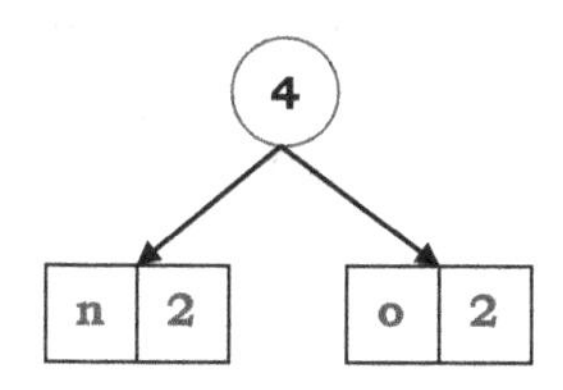

Queue will have C = {s, t, r, e, a, z} **F(z) = 4**

F(C) = {5, 5, 4, 5, 3, 4}

Now

Z = allocate a memory   X = Extract_Min(Q) = a|3

Y = Extract_Min(Q) = r|4

Left [Z] = X   Right[Z] = Y   f(Z) ← f(X) + f(Y)

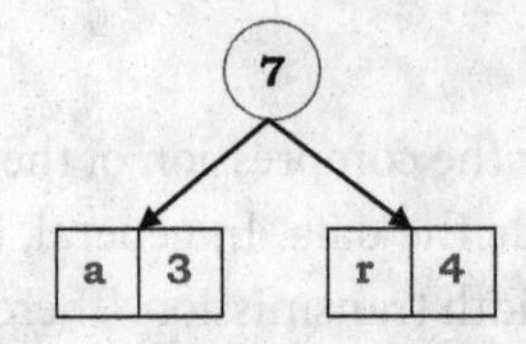

Queue will have C = {s, t, z, e, z} **F(z) = 7 and F(z) = 4**

F(C) = {5, 5, 7, 5, 4}

Now

Z = allocate a memory   X = Extract_Min(Q) = z|4

Y = Extract_Min(Q) = e|5

Left [Z] = X   Right[Z] = Y   f(Z) ← f(X) + f(Y)

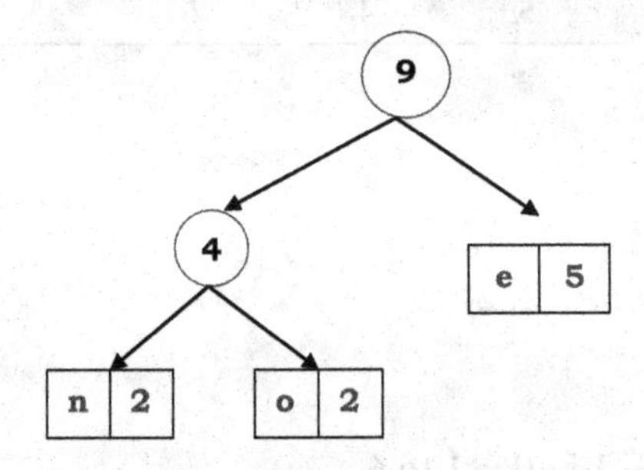

Queue will have C = {s, t, z, z} **F(z) = 7 and F(z) = 9**

F(C) = {5, 5, 7, 9}

Now

Z = allocate a memory   X = Extract_Min(Q) = s|5

Y = Extract_Min(Q) = t|5

Left [Z] = X Right[Z] = Y   f(Z) ← f(X) + f(Y)

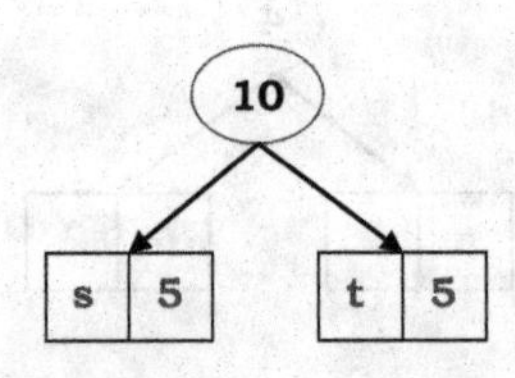

Queue will have C = {z, z, z} **F(z) = 10, F(z) = 7 and F(z) = 9**

F(C) = {10, 7, 9}

Now

Z = allocate a memory   X = Extract_Min(Q) = z|7

Y = Extract_Min(Q) =z|9

Left [Z] = X Right[Z] = Y f(Z) ← f(X) + f(Y)

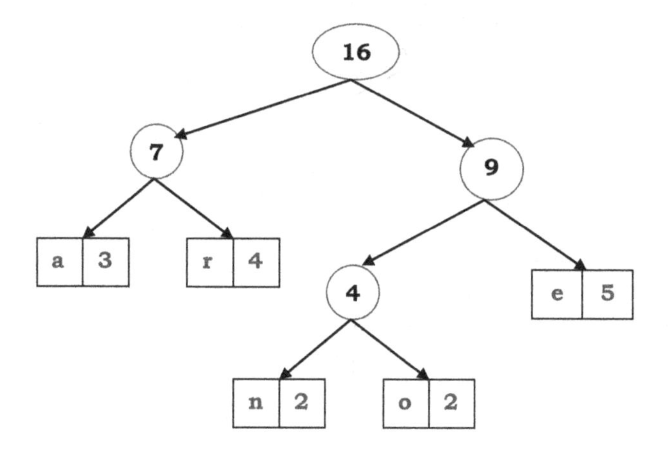

Queue will have C = {z, z} **F(z) = 10, F(z) = 16**
F(C) = {10, 16}
Now
Z = allocate a memory X = Extract_Min(Q) = z|10
Y = Extract_Min(Q) =z|16
Left [Z] = X Right[Z] = Y f(Z) ← f(X) + f(Y)

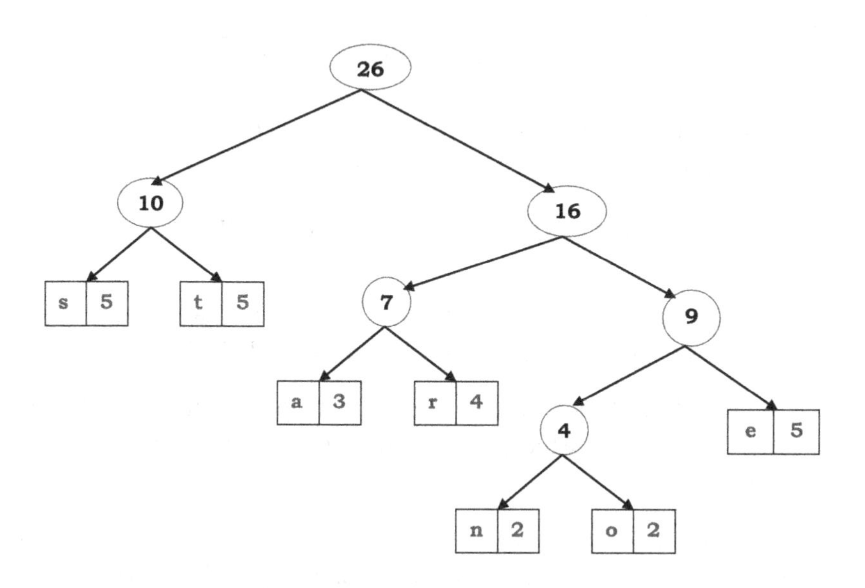

Now for encoding label 0 for the left edge and 1 for the right edge

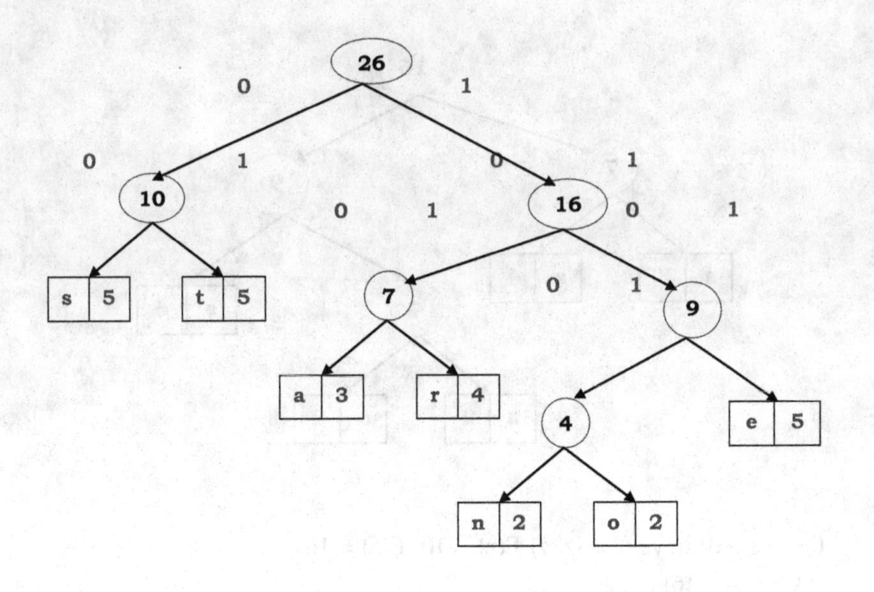

Code of

Character	s	t	r	e	a	o	n
Frequency	00	01	101	111	100	1101	1100

Therefore, the encoded string for "streets are stone stars are not" (using variable length) will be

00-01-101-111-111-01-00 100-101-111 00-01-1101-1100-111 00-01-100-101-00 100-101-111 1100-1101-01

5.5 GREEDY VERSUS DYNAMIC PROGRAMMING

- Both techniques are optimization techniques, and both build solutions from a collection of choices of individual elements.
- The greedy method computes its solution by making its choices in a serial forward fashion, never looking back or revising previous choices.
- Dynamic programming computes its solution bottom up by synthesizing them from smaller subsolutions, and by trying many possibilities and choices before it arrives at the optimal set of choices.

- There is no a priori litmus test by which one can tell if the greedy method will lead to an optimal solution.
- By contrast, there is a litmus test for dynamic programming, called the Principle of Optimality.

5.6 DATA STRUCTURES FOR DISJOINT SETS

A disjoint set data structure is a collection of sets $S = \{S_1, S_2, \ldots, S_k\}$ such that $Si \cap Sj = \{\phi\}$ for $i \neq j$.

Each set is identified by an element called the representative of the set.

A **disjoint-set data structure** is a data structure that keeps track of a set of elements partitioned into a number of disjoint (no overlapping) subsets. A **union-find algorithm** is an algorithm that performs two useful operations on such a data structure:

- *Find*: Determine which subset a particular element is in. This can be used for determining if two elements are in the same subset.
- *Union*: Join two subsets into a single subset.

Because it supports these two operations, a disjoint-set data structure is sometimes called a *union-find data structure* or *merge-find set*. The other important operation, *MakeSet*, which makes a set containing only a given element (a singleton), is generally trivial. With these three operations, many practical partitioning problems can be solved.

There is no rule that which member is used as the representative. The number of representative may be chosen depending upon certain criteria.

In disjoint set, we wish to do the following operations:

1. **MAKE-SET(x)**: This operation is used to create a new set whose only member is pointed to by – x. Since the set are disjoint, we require that 'x' not already be in some other set.
2. **UNION(x, y)**: This operation unities the dynamic sets that contain x and y say S_x and S_y into a new set that is the union of these two sets. Here, S_x and S_y are disjoint sets initially.
3. **FIND-SET(x)**: This operation returns a pointer to the representative of the set containing – x

procedure makeset(x)

```
π(x) = x
rank(x) = 0
```

function find(x)

```
while x ≠ π (x)
 x = π (x)
return x
```

5.6.1 Application of Disjoint Set Data Structures

The important applications of disjoint set data structures are to find the connected components of an undirected graph.

Algorithm Connected-Components(G)

```
1. For each vertex v ∈ V(G)
2. do MAKE-SET(v)
3. for each edge <u,v> ∈ E(G)
4. do if FIND-SET(u) ≠ FIND-SET(v)
5. Then UNION(u,v)
```

SAME – COMPONENT(u,v): [It determine whether two vertices are in the same connected component or not]

If FIND-SET(u) = FIND – SET (v)
Then return TRUE
Else
Return FALSE

Example:

A graph with four connected components as (Figures 5.1–5.4)

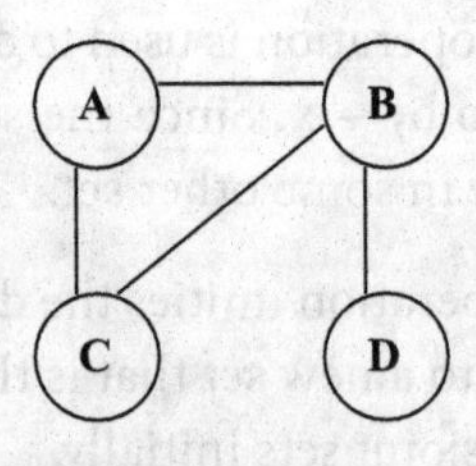

FIGURE 5.1 Four connected components.

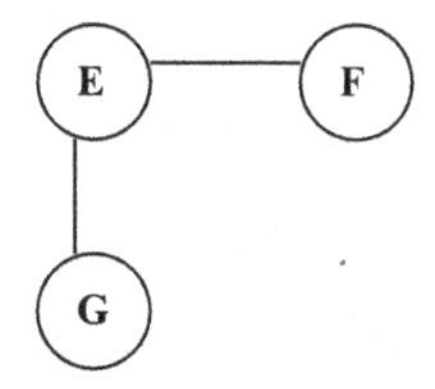

FIGURE 5.2 Three connected components.

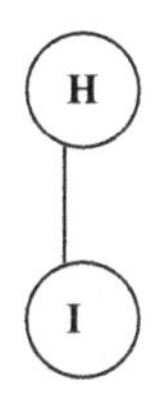

FIGURE 5.3 Two connected components.

FIGURE 5.4 Single component graph.

{A, B, C, D}, {E, F, G}, {H, I}, {J}
Apply the above algorithm
E = {(B, D), (E, G), (A, C), (H, I), (A, B), (E, F), (B, C)}

1. Make a set of each vertex that is
   V = {{A}, {B}, {C}, {D}, {E}, {F}, {G}, {H}, {I}, {J}}
2. For each edge in E do
   UNION (B, D)
   V = {{A}, {B, D}, {C}, {E}, {F}, {G}, {H}, {I}, {J}}
   UNION (E, G)
   V = {{A}, {B, D}, {C}, {E, G}, {F}, {H}, {I}, {J}}
   UNION(A, C)
   V = {{A, C}, {B, D}, {E, G}, {F}, {H}, {I}, {J}}
   UNION(H, I)
   V = {{A, C}, {B, D}, {E, G}, {F}, {H, I}, {J}}

UNION(A, B)
V = {{A, B, C, D}, {E, G}, {F}, {H, I}, {J}}
UNION(E, F)
V = {{A, B, C, D}, {E, F, G}, {H, I}, {J}}

For the edge {B, C}, there is no UNION operation because both the vertices belong to the same component.

Therefore, final connected components are V = {{A, B, C, D}, {E, F, G}, {H, I}, {J}}.

5.6.2 Linked List Representation of Disjoint Sets

We can represent disjoint sets by using a linked list. In a linked list, representation of a disjoint set first object in each linked list serves as its sets representative. Each object in the linked list contains a set member, a pointer to the object containing the next set member, and a pointer back to the representative.

Each list maintains pointers head to the representative and tail to the last object in the list.

The linked list representation takes O(1) time for both MAKE-SET and FIND-SET operations, whereas UNION operation takes

$$\sum_{i=1}^{n-1} i = \theta\left(n^2\right)$$

Example:

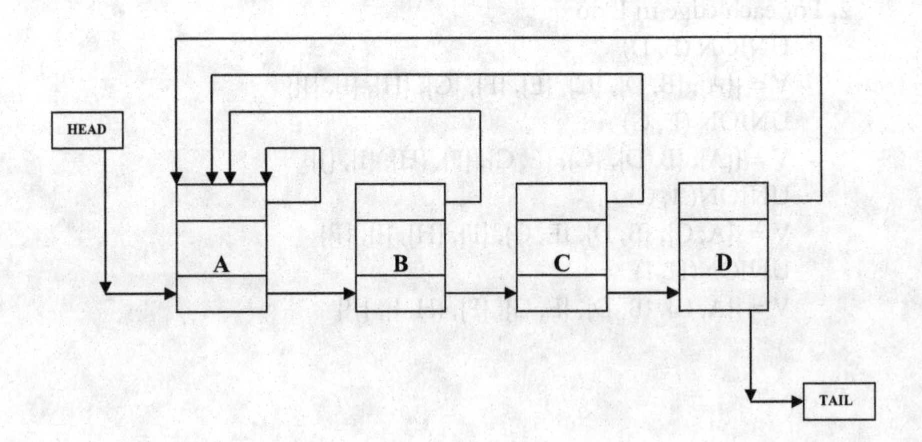

5.6.3 Disjoint Set of Forests

In this MAKE-SET operation, simply create a tree with just one node. We perform a FIND-SET operation by choosing a parent pointer until we find the root of the tree. In the UNION operation, the root of one tree points to the root of another tree.

Here we apply UNION by RANK, i.e., we find the size of a smaller tree then it is pointed by the root of the largest tree.

- **Union by Rank**: When linking two trees, make root with smaller rank a child of the root with a larger rank.

  Another method is PATH COMPRESSION where each node is directly pointing to the root.

- **Path Compression**: When executing FIND(x), make all nodes on the find path of x direct children of the root.

Algorithms for Disjoint Forests Union by Rank

Properties

Assuming only union-by-rank, the following three properties hold:

Property 1: For any non-root node x, rank(x) < rank(p(x)).

Property 2: The subtree rooted at a rank-k node has at least 2^k nodes.

Since a node of rank k is created by linking two nodes of rank k – 1, Property 2 holds by induction on k. Properties 1 and 2 entail.

Property 3: The number of nodes of rank k is at most $n/2^k$.

MAKE-SET(x)

```
1. P[x] ← 0
2. Rank[x] ← 0
```

UNION(x, y)

```
1. LINK (FIND-SET(x), FIND-SET(y))
```

LINK(x, y)

```
1. If Rank[x] > Rank[y]
2. then P[y] ← x
3. else
4. P[x] ← y
5. If Rank[x] = Rank[y]
6. then Rank[y] ← Rank[y]+1
```

FIND-SET(x)

```
1. If x ≠ P[x]
2. then P[x] ← FIND-SET(P[x])
3. Return P[x]
```

The FIND-SET procedure is a two-pass method: it makes one pass up the find path to find the root and it makes a second pass back down the find path to update each node so that it points directly to the root.

Note: For all tree roots x, SIZE[x] ≥ $2^{RANK[X]}$

Example:

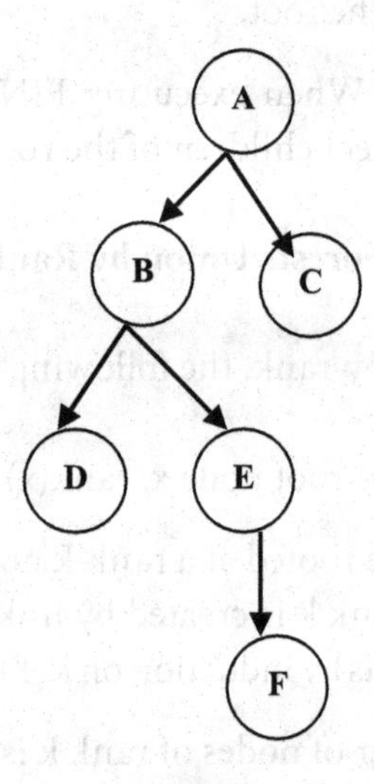

FIGURE 5.5

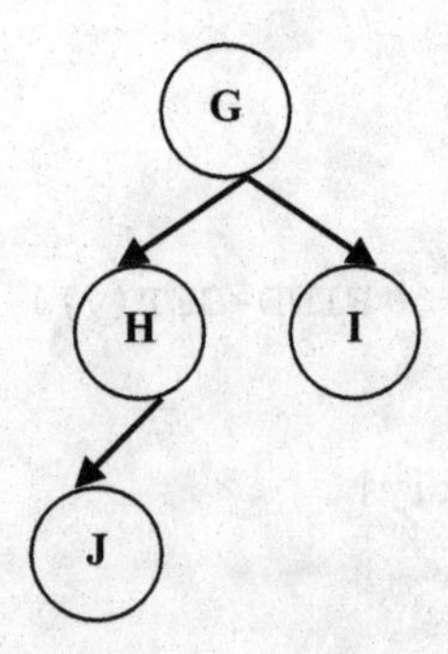

FIGURE 5.6

Rank of Figure 5.5 is 3

Rank of Figure 5.5 is 2

Therefore, Figure 5.6 will be attached with the parent of Figure 5.5 in the concept of Union by Rank.

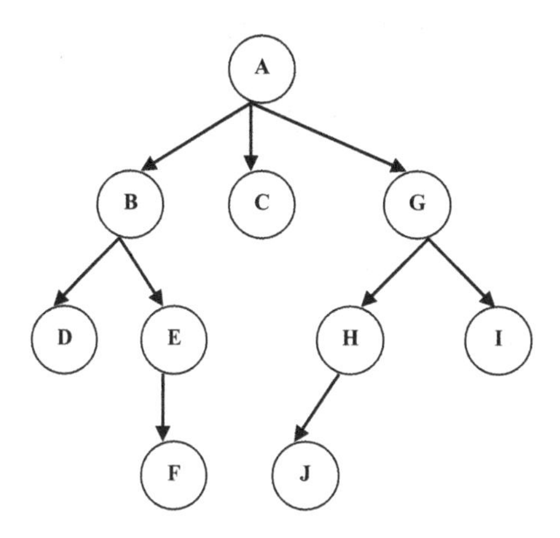

5.6.4 By Path Compression

During find(i), as we traverse the path from i to root, update parent entries for all these nodes to the root. This reduces the heights of all these nodes.

Updated code for find

```
find (i) {
if (parent[i] < 0)
return i;
else return parent[i] = find (parent[i]);
}
```

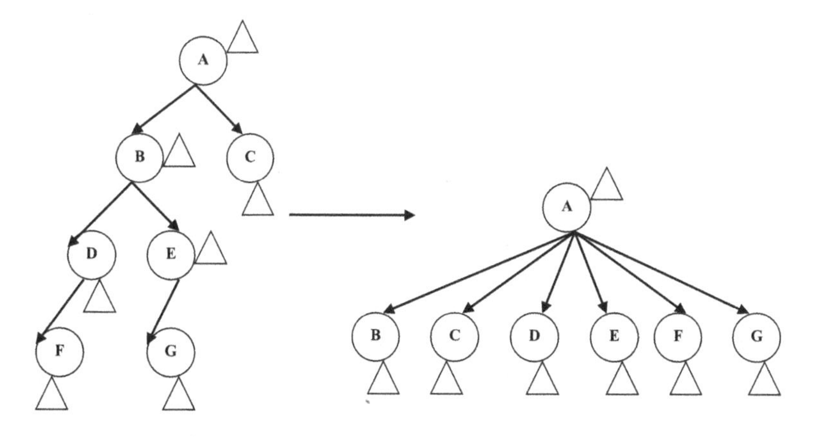

Kruskal Algorithm an Application

MST-KRUSKAL(G, w)

```
1. A ← ϕ
2. for each vertex v ∈ V [G]
3. do MAKE-SET(v)
4. sort the edges of E into nondecreasing order by
   weight w
5. for each edge (u, v) ∈ E, taken in nondecreasing
   order by weight
6. do if FIND(u) 6= FIND(v)
7. then A ← A ∪ {(u, v)}
8. UNION(u, v)
9. return A
```

Example:

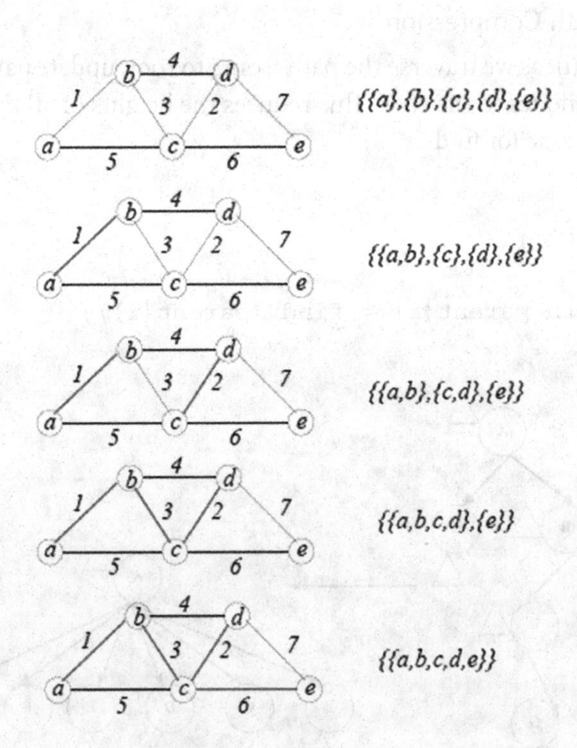

5.7 QUESTIONS

5.7.1 Short Questions

1. Write the basic structure of the greedy algorithm.
2. What is the knapsack problem?
3. What is HUFFMAN ENCODING?
4. Why greedy algorithm is used?
5. How does Kruskal's algorithm know when the addition of an edge will generate a cycle?
6. What is feasibility?
7. Define how the greedy algorithm optimal.
8. How many types of criteria solved the knapsack problem?
9. Define fractional knapsack.
10. Define encoding and decoding.

5.7.2 Long Questions

1. Write the ACTIVITY-SCHEDULE algorithm.
2. Write the knapsack algorithm.
3. Short notes on greedy and dynamic programming.
4. Explain how linked list representation in disjoint sets with proper example.
5. Write algorithm for disjoint forests union by rank.
6. Write the fractional knapsack algorithm and solve the problem.
   I = {i1, i2, i3, i4, i5}
   W = {5, 10, 20, 30, 40}
   V = {30, 20, 100, 90, 160}
7. Write the HUFFMAN Algorithm and find an optimal Huffman code for the following A = 50 B = 25 C = 15 D = 40 E = 75

8. Write Kruskal's algorithm with an example.
9. Construct the Huffman tree

g , 6	d , 13	f , 17	b,18	c ,29	e ,30	a , 37

10. Write algorithm for connected components with an example.

CHAPTER 6

Graph

6.1 TRAVERSAL OF GRAPH

The Graph Traversal is of two types:

- Breadth First Search (BFS)
- Depth First Search (DFS).

6.1.1 Breadth First Search

BFS starts at a given vertex, which is at level '0'. In the first stage, we visit all vertices at level – 1. In the second stage, we visit all vertices at level 2.

These new vertices, which are adjacent to level 1 vertices and so on.

The BFS terminates when every vertex has been visited.

BFS used to solve

1. Testing whether or not the graph is connected.
2. Computing a spanning forest of Graph.
3. Computing a cycle on graph or reporting that no such cycle exists.
4. Computing for every vertex on graph, a path with the minimum number of edges between the start vertex and the current vertex or reporting that no such path exists.

DOI: 10.1201/9781003093886-6

Analysis: Total running time of BFS = O(V + E)

Algorithm BFS(G, S)

```
 1. for each vertex u ∈ V[G]- {S}
 2. do color[u] ← white
 3. d[u] ← ∞ i.e./distance from S
 4. P[u] ← NIL i.e./Parent in the BFS tree
 5. color[S] ← gray
 6. d[S] ← 0
 7. Q ← {S}
 8. while Q φ do
 9. u ← head[Q]
10. for each v ∈ Adj[u] do
11. if color[v] = white then
12. color[v] ← gray
13. d[v] ← d[u] + 1
14. p[v] ← u
15. ENQUEUE(Q,v)
16. DEQUEUE(Q)
17. color[u] ← black
```

Example:

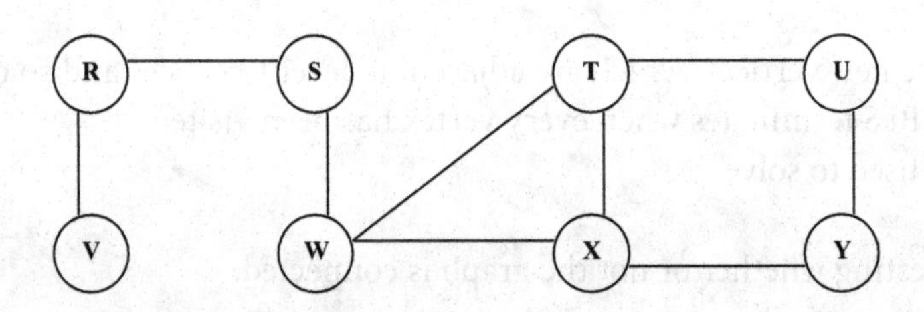


```
STEP 1 BFS(G,S)
for each vertex u ∈V[G]- {S}
    do color[u] ← white
              so color[S] = white
              color[S] = gray
    d[S] = 0
    Q = {S}
```

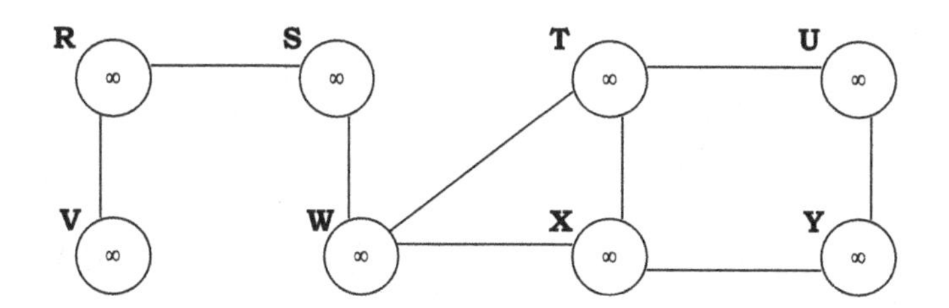


```
STEP 2 While Q ≠ ϕ
      Since Q is not empty so S = Head[Q]
for each v ∈ Adj[u]
i.e/ for each v ∈ Adj[S]
find Adj[S] = [R,W]
if color[v] = white i.e./color[R] = WHITE and color[W]
=  white
so color[R] = gray and color[W] = gray
 d[v] = d[u] + 1 i.e./d[R] = d[S] + 1 = 0 + 1 = 1
                      d[W] = d[S] + 1 = 0 + 1 = 1
ENQUE(Q,v)
i.e/ ENQUE(Q,R) and ENQUE(Q,W)
DEQUEUE(Q) and color[S] = BLACK
    So DEQUEUE(S) and color[S] = black.
```

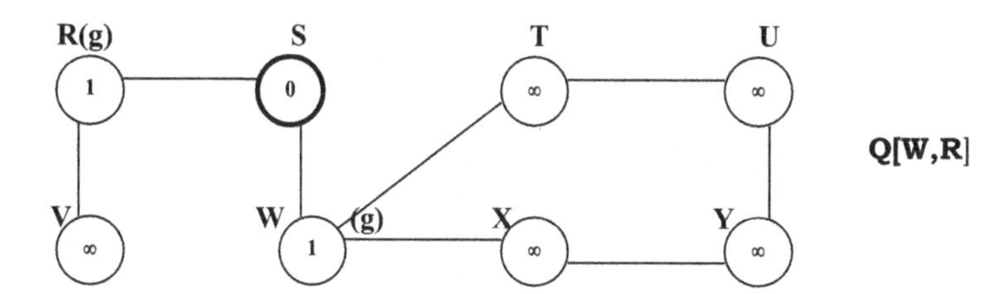

STEP 3

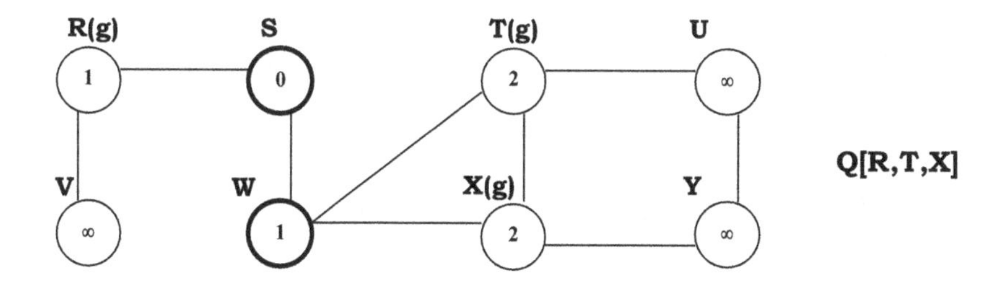

STEP 4

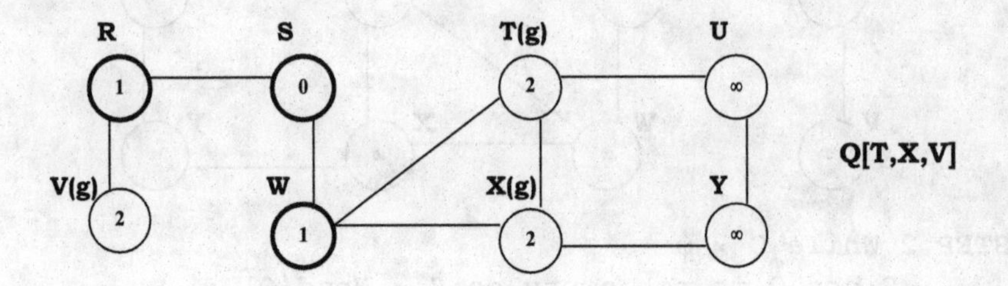

STEP 5

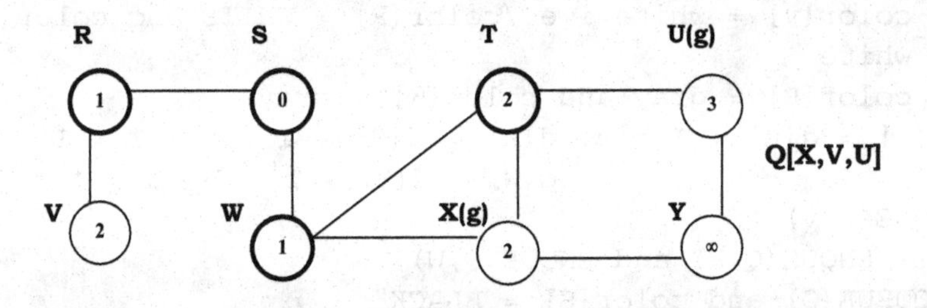

STEP 6

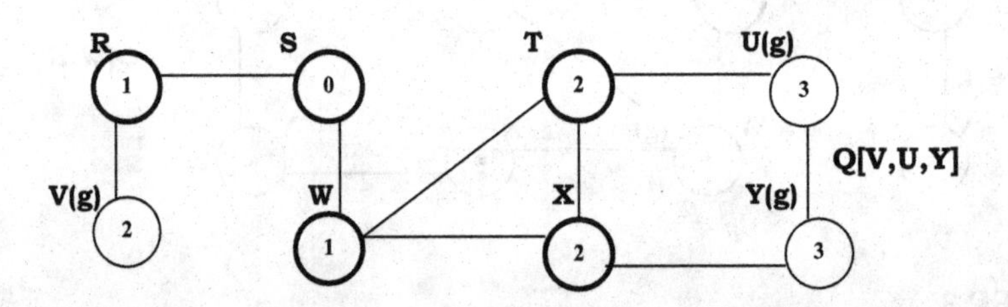

STEP 7

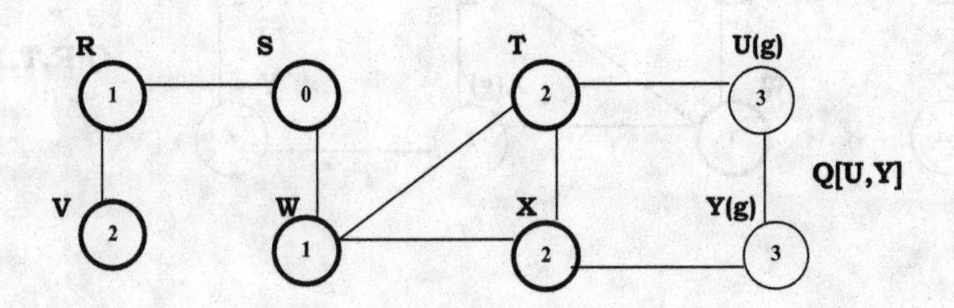

STEP 8

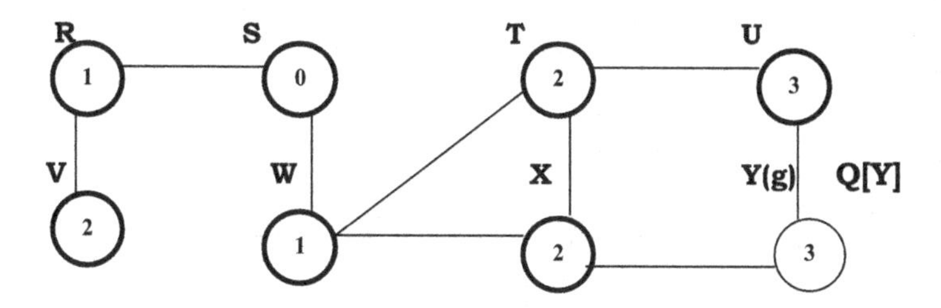

STEP 9

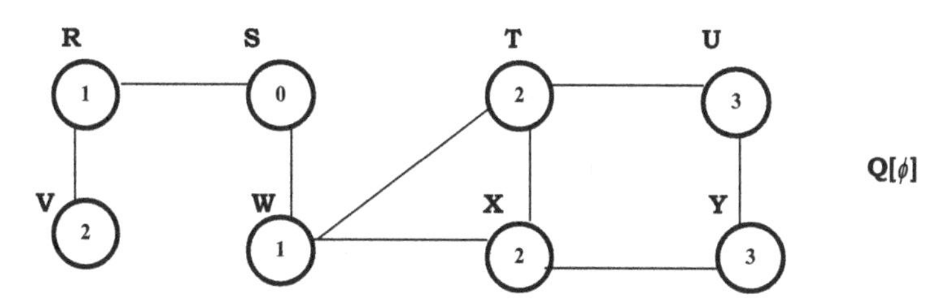

The result of BFS is **S, W, R, T, X, V, U, Y.**

6.1.2 Depth First Search

Depth First Search is another way of traversing of graph. It uses a STACK data structure for traversing.

Algorithm DFS(G)

```
1. for each vertex u ∈ V[G]
2. do color[u] ← white
3. π(u) ← NIL
4. time ← 0
5. for each vertex u ∈ V[G]
6. do if color[u] ← white
7. DFS-VISIT(u)
```

Algorithm DFS-VISIT(u)

```
1. color[u] ← gray
2. time ← time + 1
3. d[u] ← time
4. for each v ∈ Adj[u]
5. if  color[v] ← white
6. then π(v) ← u
```

```
7. DFS-VISIT(v)
8. color(u) ← black
9. finish[u] ← time ← time + 1
```

Example:

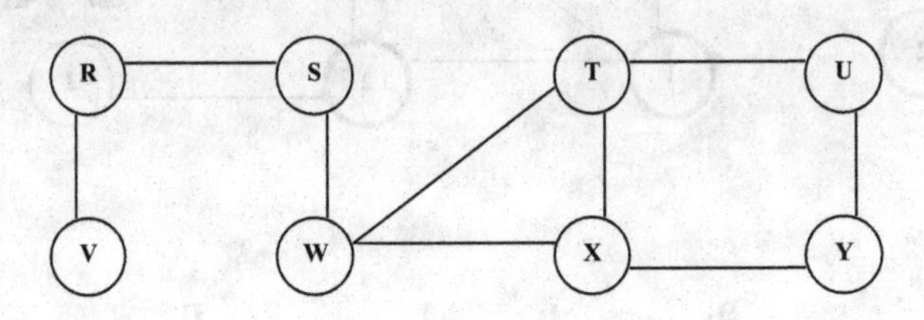

STEP 1 u = R

Color[R] = gray, time = 0 + 1 = 1 d[R] = 1

$$v \in \text{Adj}[u] \Rightarrow v = (S, V)$$

If color(S) = white and since color(S) = white

$$\pi(V) = u \Rightarrow \pi(s) = R$$

COLOR
PARENT (π)

R	S	T	U	V	W	X	Y
Nil	Nil	Nil	Nil	Nil	Nil	Nil	Nil
	R	**W**	**Y**	**R**	**S**	**T**	**X**

TIME

R	S	T	U	V	W	X	Y
1	2	4	7	14	3	5	6
16	**13**	**11**	**8**	**15**	**12**	**10**	**9**

6.2 SPANNING TREE

Let a graph G = (V, E), if T is a subgraph of G and contains all the vertices but no cycles/circuit, then T may be called a Spanning Tree.

6.2.1 Minimum Spanning Tree

If a weighted graph is considered, then the weight of the spanning tree (T) of graph G can be calculated by summing all the individual weights, in the spanning tree T. However, we observe that for a graph, a number of spanning trees are available, but a minimum spanning tree means the spanning tree with a minimum weight.

A tree is a connected graph with no cycles.

1. A graph is a tree if and only if there is one and only one path joining any two of its vertices.
2. A connected graph is a tree if and only if every one of its edges is a bridge.
3. A connected graph is a tree if and only if it has N vertices and N – 1 edges.

One practical implementation of MST would be in the design of network.

Another useful application of MST is to find airline routes.

6.2.2 Kruskal Algorithm

The Kruskal algorithm is used to build the minimum spanning tree in forest. Initially, each vertex is in its own tree in forest. Then algorithm considers each edge, in turn, order by increasing weight. If an edge (u, v) connects two different trees, then (u, v) is added to the set of edges of MST, and two trees connected by an edge (u, v) are merged into a single tree.

On the other hand, if an edge (u, v) connects two vertices in the same tree, then edge (u, v) is discarded. It uses a disjoint set data structure to maintain several disjoint sets of elements.

Each set contains the vertices in a tree of the current forest.

Algorithm kruskal (G, W)

```
1. A = {ϕ}
2. for each vertex v ∈ V[G]
3. do MAKE-SET(v)
4. Sort edges E by increasing order of weight W
5. for each edge(u,v) in E(G)
```

```
6. do if FIND-SET(u) ≠ FIND-SET(v)
7. then A = A  ∪ {(u,v)}
8. UNION(u,v)
9. return A
```

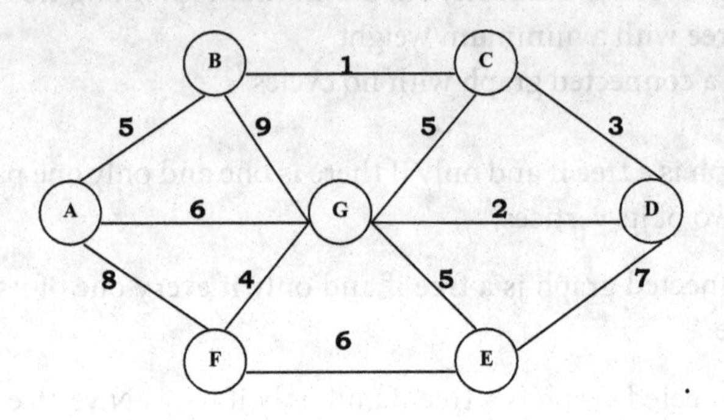

Example:

First, sort the edges according to their weights in ascending order.

Edges	Weight
(B, C)	1
(G, D)	2
(C, D)	3
(F, G)	4
(A, B)	5
(C, G)	5
(G, E)	5
(A, G)	6
(F, E)	6
(E, D)	7
(A, F)	8
(B, G)	9

Now connect every edge from the beginning of the above list, and if a closed path is found, then discard that edge.

(B, C)

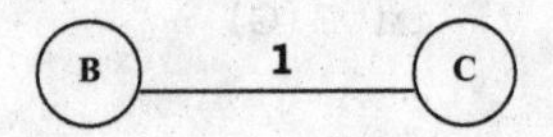

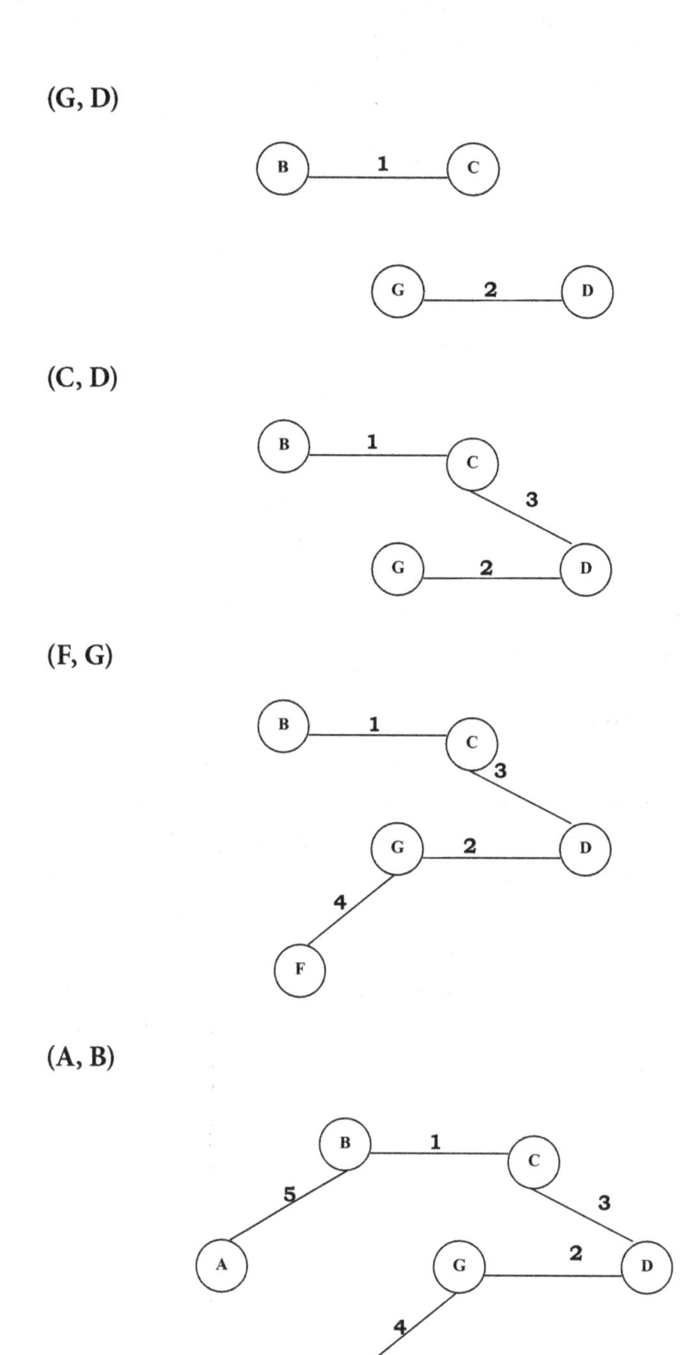

Since (C, G) forms a closed path, so discard it.

(G, E)

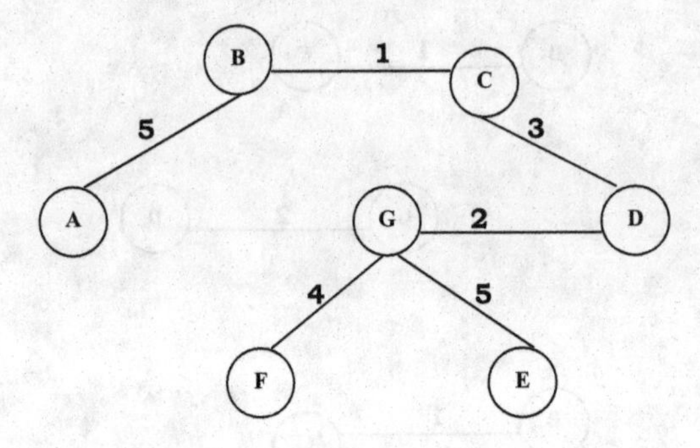

Since (A, G), (F, E), (E, D), (A, F) and (B, G) forms closed paths so discard them and the above graph is the minimum spanning tree.

The Kruskal algorithm is a greedy algorithm because at each step it adds to the forest an edge at least possible.

6.2.3 Prim's Algorithm

The key to implementing Prim's algorithm efficiently is to make it easy to select a new edge to be added to the tree formed by the edges in A in the pseudo-code below, the connected graph G and the root R of the minimum spanning tree to be grown are inputs to the algorithm. During the execution of the algorithm, all vertices that are not in the tree reside in a min priority queue Q based on a key field. For each vertex V, Key[V] is the minimum weight of any edge connecting V to a vertex in the tree, by conversion key[V] = ∞ if there is no such edge. The field π (V) names the parent of V in the tree.

Algorithm Prim (G, W, R)

```
 1. for each u ∈ V(G)
 2. do key[u] ← ∞
 3. π (u) ← Nil
 4. Key[R] ← 0
 5. Q ← V(G)
 6. While Q ≠  ϕ
 7. do u ← EXTRACT-MIN(Q)
 8. for each v ∈ Q and w(u,v) < key(v)
 9. then π (v) ← u
10. key[R] ← w(u,v)
```

Example:

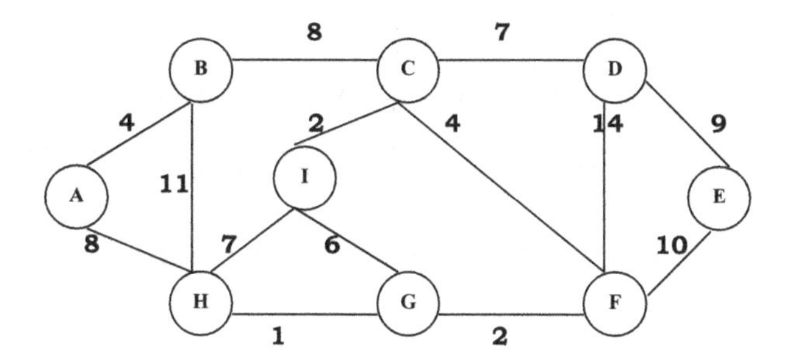

Queue:

A	B	C	D	E	F	G	H	I

KEY

Steps	A	B	C	D	E	F	G	H	I
	∞	∞	∞	∞	∞	∞	∞	∞	∞
0	0								
1		4						8	
2			8						
3				7		4			2
4							6	7	
5					10		2		
6								1	
7					9				

PARENT (π)

Steps	A	B	C	D	E	F	G	H	I
	Nil	Nil	Nil	Nil	Nil	Nil	Nil	Nil	Nil
1		A						A	
2			B						
3				C		C			C
4							I	I	
5					F		F		
6								G	
7					D				

Step 1: u = Extract_min(Q) i.e./A and Key[A] = 0

V = {B, H} [Adjacent of A]

$W(A,B) < Key[B]$ i.e./ $4 < \infty$ so **Key[B] = 4** **π(B) = A**

$W(A,B) < Key[H]$ i.e./ $8 < \infty$ so **Key[H]** = 8 **π(H) = A Delete A**

Step 2: u = Extract_min(Q) i.e./B because the Minimum of [4, 8] so 4 i.e./B

V = {C,H} [Adjacent of B]

$W(B,C) < Key[C]$ i.e./ $8 < \infty$ so **Key[C] = 8** **π(C) = B**

W(B,H) < Key[H] i.e./8 < 8 Since condition is false so no action will take

Delete: B

Step 3: u = Extract_min(Q) i.e./Min of (8, 8) so C

V = {B, I, F, D} [Adjacent of C] since **B** is not in Q so **B** will not be considered.

$W(C,I) < Key[I]$ i.e./ $2 < \infty$ so **Key[I]** = 2 **π(I) = C**

$W(C,F) < Key[F]$ i.e./ $4 < \infty$ so **Key[F] = 4** **π(F) = C**

$W(C,D) < Key[D]$ i.e./ $7 < \infty$ so **Key[D] = 7** **π(D) = C Delete C**

Step 4: u = Extract_min(Q) i.e/ Min of (7,4,8,2) so I

V = {C, H, G} [Adjacent of I] Since **C** is not in Q so **C** will not be considered

W(I, H) < Key[H] i.e./7 < 8 so **Key[H]** = 7 **π(H) = I**

W(I, G) < Key[G] i.e./6 < ∞ so **Key[G] = 6** **π(G) = I** Delete **I**

Step 5: u = Extract_min(Q) i.e./ Min of (7, **4**, 6, 7) so F

V = {E, C, D, G} [Adjacent of F] since **C** is not in Q so **C** will not be considered

W(F, G) < Key[G] i.e./2 < 6 so **Key[G] = 2** **π(G) = F**

W(F, E) < Key[E] i.e./10 < ∞ so **Key[E] = 10** π(E) = **F Delete F**

W(F, D) < Key[D] i.e./14 < 7 (False)

Step 6: u = Extract_min(Q) i.e./Min of (D, E, G, H) (7, 10, 2, 7) so G

V = {H, I, F} [Adjacent of G] **I, F** will not be considered (Not in Queue)

W(G, H) < Key[H] i.e./1 < 7 so **Key[H] = 1** **π(H) = G Delete G**

Step 7: u = Extract_min(Q) i.e./Min of (D, E, H) (7, 10, 1) so H

V = {A, B, I, G} [Adjacent of H]

Since all are not in Queue so all will be discarded

Delete H

Step 8: u = Extract_min(Q) i.e./Min of (D, E) (7, 10) so D

V = {C, F, E} [Adjacent of D] **C, F** will not be considered (Not in Queue)

W(D, E) < Key[E] i.e./9 < 10 so **Key[E] = 9** **π(E) = D Delete D**

Step 9: u = Extract_min(Q) i.e./Min of (E) (9) so E

V = {D, F} [Adjacent of H]

Since all are not in Queue so all will be discarded

Delete E

Finally, all the vertices are deleted from the Q. Now plot the graph according to the parent table.

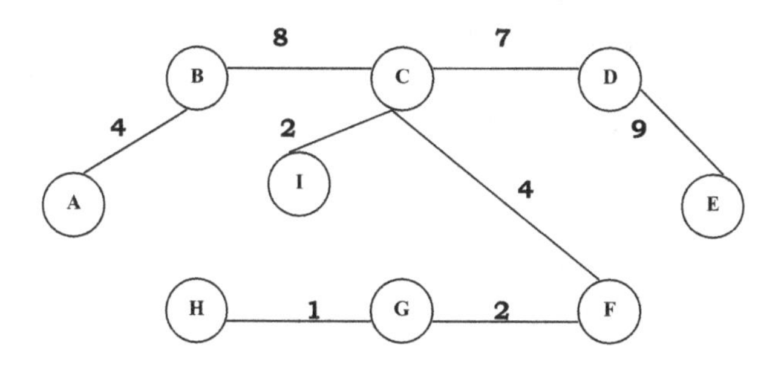

6.3 SINGLE-SOURCE SHORTEST PATH

The shortest path weight from a vertex $u \in V$ to a vertex $v \in V$ in the weighted graph is the minimum cost of all paths from u to v if there exists no such path from vertex u to vertex v then the weight of the shortest path is ∞.

We can also define it as

$$\delta(u,v) = \begin{cases} \text{Min}\left\{w(p): u \xrightarrow{p} v \text{ if there is a path from u to v}\right\} \\ \infty \qquad\qquad\qquad \text{otherwise} \end{cases}$$

6.3.1 Negative Weighted Edges

The negative weight cycle is a cycle whose total weight is –ve. No path from starting vertex S to a vertex on the cycle can be the shortest path.

Since a path can run around the cycle, many times so it may get any –ve costing other word we can say that a negative cycle invalidates the notion of distance based on edge weights.

If some path from S to v contains a negative cost style, there does not exist the shortest path; otherwise, there exists one that is simple.

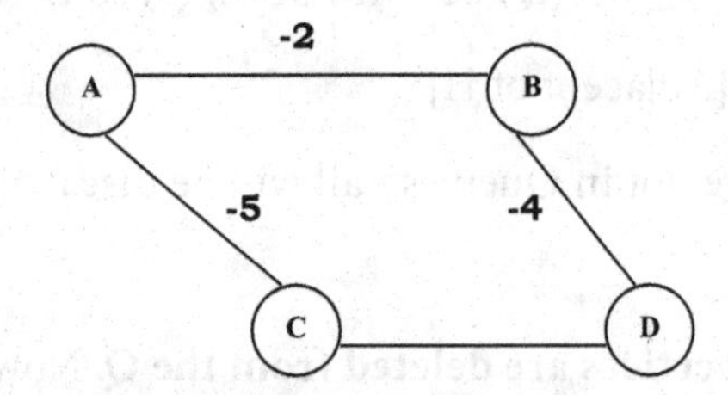

6.3.2 Relaxation Technique

This technique consists of testing whether we can improve the shortest path found so far; if so, update the shortest path. A relaxation step may or may not decrease the value of the shortest path estimate.

Algorithm RELAX(u, v, w)

```
1. if d[u] + w(u,v) <d[v]
2. then d[v] ← d[u] + w(u,v)
3. π[v] ← u
```

Example:

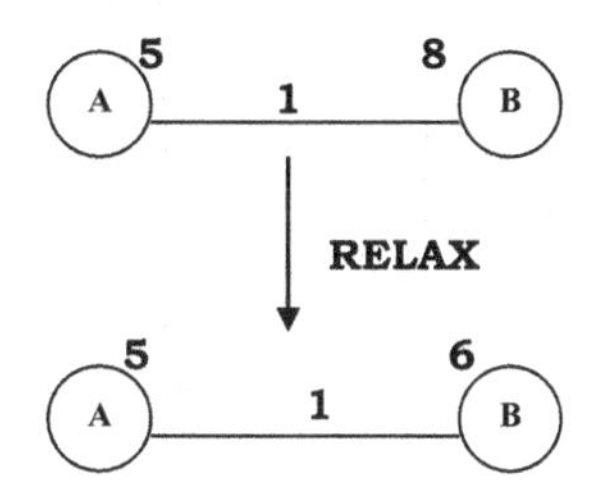

Algorithm INITIALIZE_SINGLE_SOURCE(G, S)

```
1. for each vertex v ∈ V[G]
2. do d[v] ← ∞
3. π(v) ← Nil
4. d[s] ← 0
```

6.3.3 Bellman–Ford Algorithm

Bellman–Ford algorithm solves the single-source shortest path problem in the general case in which edges of a given digraph can have –ve weight as long as G contains no negative cycles.

It uses d[u] as an upper bound on the distance d[u, v] from u to v. The algorithm progressively decreases an estimate d[v] on the weight of the shortest path from the source vertex S each vertex v in V until it achieves the actual shortest path.

This algorithm returns TRUE if the given digraph contains no –ve cycle that is reachable from source vertex S, otherwise FALSE.

Algorithm BELLMAN–FORD(G, W, S)

```
1. INITIALIZE-SINGLE-SOURCE(G,S)
2. for each vertex i = 1 to V[G] - 1 do
3. for each edge(u,v) ∈ E(G) do
4. RELAX(u,v,w)
5. for each edge(u,v) in E(G)
6. do if d[u] + w(u,v) < d[v]
7. then return FALSE
8. Return TRUE
```

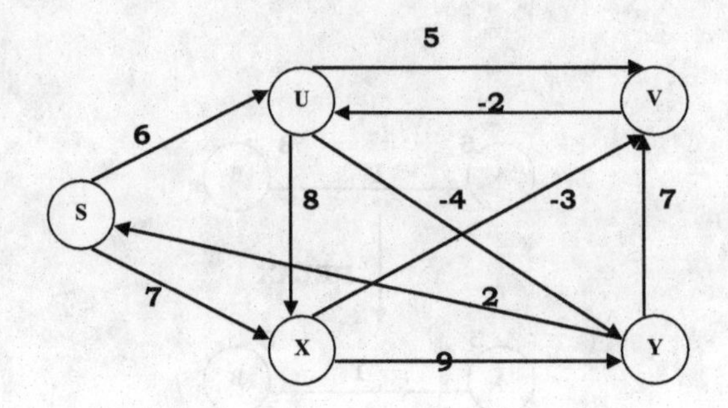

Distance

Steps	S	U	V	Y	X
	∞	∞	∞	∞	∞
	0				
1		6			7
2			11	2	
3			9		
4			4		
5		2			
6				−2	
7	No effect				

Parent (π)

Steps	S	U	V	Y	X
	Nil	Nil	Nil	Nil	Nil
1		S			S
2			U	U	
3			Y		
4			X		
5		V			
6				U	
7	No effect				

STEPS

For i = 1: Consider the vertex S

Now Since the Adj(S) = (U, X) now implement Relax with (S, U) and (S, X)

Relax(s, u, w)
$d[s] + w(s, u) < d[u] = 0 + 6 < \infty$ so $d[u] = 6$ and $\pi(u) = S$

Relax(s, x, w)
$d[s] + w(s, x) < d[x] = 0 + 7 < \infty$ so $d[x] = 7$ and $\pi(x) = S$

For i = 2: Consider the vertex U (min(u, x))
Now Since the Adj(U) = (V, X, Y) now implement Relax with (U, V), (U, X) and (U, Y)

Relax(u, v, w)
$d[u] + w(u, v) < d[v] = 6 + 5 < \infty$ so $d[v] = 11$ and $\pi(v) = u$

Relax(u, x, w)
$d[u] + w(u, x) < d[x] = 6 + 8 < 7$ (False)

Relax(u, y, w)
$d[u] + w(u,y) < d[y] = 6 + -4 < \infty$ so $d[y] = 2$ and $\pi(y) = u$

For i = 3: Consider the vertex Y (minimum)
Now Since the Adj(Y) = (V, S) now implement Relax with (Y, V) and (Y, S)

Relax(y, v, w)
$d[y] + w(y, v) < d[v] = 2 + 7 < 11$ so $d[v] = 9$ and $\pi(v) = y$

Relax(y, s, w)
$d[y] + w(y, s) < d[s] = 2 + 2 < 0$ (False)

For i = 4: Consider the vertex X (minimum)
Now Since the Adj(X) = (V, Y) now implement Relax with (X, V) and (X, Y)

Relax(x, v, w)
$d[x] + w(x, v) < d[v] = 7 + -3 < 9$ so $d[v] = 4$ and $\pi(v) = x$

Relax(x, y, w)
$d[x] + w(x, y) < d[y] = 7 + 9 < 2$ (False)

For i = 5: Consider the vertex V (minimum)
Now Since the Adj(V) = (U) now implement Relax with (V, U)

Relax(v, u, w)
$d[v] + w(v, u) < d[u] = 4 + -2 < 6$ so $d[u] = 2$ and $\pi(u) = v$

For i = 6: Consider the vertex U (minimum)
Now Since the Adj(U) = (V, Y, X) now implement Relax with (U, V), (U, Y), (U, X)

Relax(u, v, w)
$d[u] + w(u, v) < d[v] = 2 + 5 < 4$ (False)

Relax(u, y, w)
$d[u] + w(u, y) < d[y] = 2 + -4 < 2$ so $d[y] = -2$ and $\pi(y) = u$

Relax(u, x, w)
$d[u] + w(u, x) < d[x] = 2 + 8 < 7$ (False)

For i = 7: Consider the vertex Y (minimum)
Now Since the Adj(Y) = (V, S) now implement Relax with (Y, V), (Y, S)

Relax(y, v, w)
$d[y] + w(y, v) < d[v] = -2 + 7 < 4$ (False)

Relax(y, s, w)
$d[y] + w(y, s) < d[s] = -2 + 2 < 0$ (False)

So the shortest path graph is

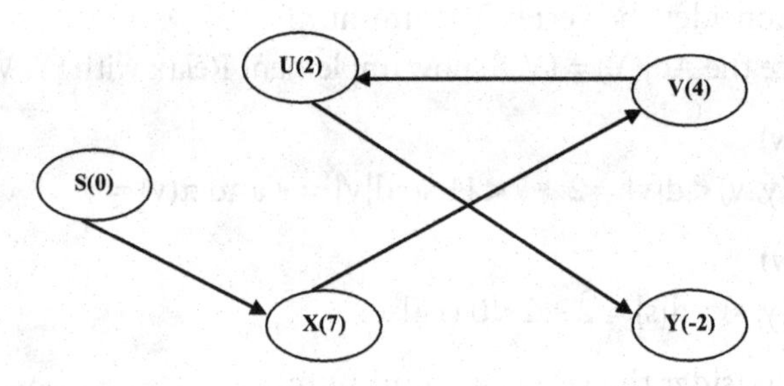

6.3.4 Dijkstra's Algorithm

Dijkstra's algorithm solves the single-source shortest path problem when all edges have non −ve weights. It is a greedy algorithm and similar to Prim's algorithm. The algorithm starts at the source vertex S it grows a tree T that ultimately spans all vertices reachable from S. Vertices are added to T in order of distance i.e./first S, then the vertex closest to S, then the next closest, and so on.

Algorithm DIJKSTRA(G, W, S)

```
1. INITIALIZE-SINGLE-SOURCE(G,S)
2. S ← { }
3. Initialize Priority Queue i.e./Q ← V[G]
4. While Q ≠ ϕ
```

```
5. do u ← Extract_min(Q)
6. S ← S ∪{u}
7. for each vertex v ∈ Adj[u]
8. do RELAX(u,v,w)
```

Example:

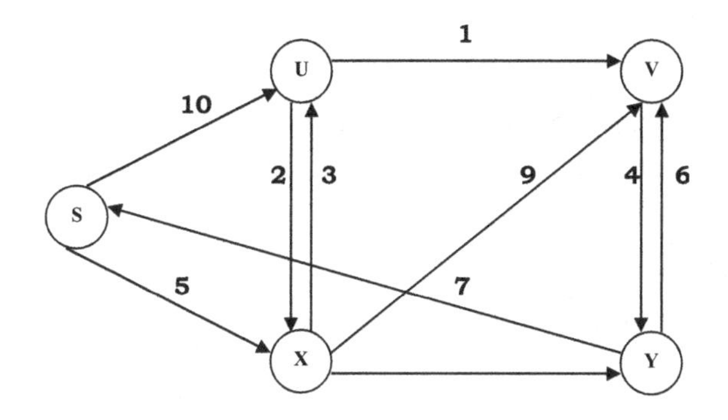

Distance

Steps	S	U	V	Y	X
	∞	∞	∞	∞	∞
	0				
1		10			5
2		8	14	7	
3			13		
4			9		

Parent (π)

Steps	S	U	V	Y	X
	Nil	Nil	Nil	Nil	Nil
1		S			S
2		X	X	X	
3			Y		
4			U		

STEP 1: Consider the vertex S

Now since the Adj(S) = (X, U) now implement Relax with (S, U) and (S, X)

Relax(s, u, w)

$d[s] + w(s, u) < d[u] = 0 + 10 < \infty$ so $d[u] = 10$ and $\pi(u) = S$

Relax(s, x, w)

$d[s] + w(s,x) < d[x] = 0 + 5 < \infty$ so $d[x] = 5$ and $\pi(x) = S$ **delete S**

STEP 2: Consider the vertex X (minimum)

Now since the Adj(X) = (U, V, Y) now implement Relax with (X, U) (X, Y) and (X, V)

Relax(x, u, w)

$d[x] + w(x, u) < d[u] = 5 + 3 < 10$ so $d[u] = 8$ and $\pi(u) = X$

Relax(x, v, w)

$d[x] + w(x, v) < d[v] = 5 + 9 < \infty$ so $d[v] = 14$ and $\pi(v) = X$

Relax(x, y, w)

$d[x] + w(x, y) < d[y] = 5 + 2 < \infty$ so $d[y] = 7$ and $\pi(y) = X$ **delete X**

STEP 3: Consider the vertex Y (minimum)

Now since the Adj(Y) = (S, V) now implement Relax with (Y, S) (Y, V)

Relax(y, s, w)

$d[y] + w(y, s) < d[s] = 7 + 7 < 0$ (False)

Relax(y, v, w)

$d[y] + w(y, v) < d[v] = 7 + 6 < 14$ so $d[v] = 13$ and $\pi(v) = Y$ **delete Y**

STEP 4: Consider the vertex U (minimum)

Now since the Adj(U) = (V, X) now implement Relax with (U, V) (U, X)

Relax(u, v, w)

$d[u] + w(u, v) < d[v] = 8 + 1 < 13$ so $d[v] = 9$ and $\pi(v) = U$

Relax(u, x, w)

$d[u] + w(u, x) < d[x] = 8 + 2 < 5$ (False) **delete U**

So final shortest path matrix is

Dsv = Dsx + Dxu + Duv = 5 + 3+ 1 = 9

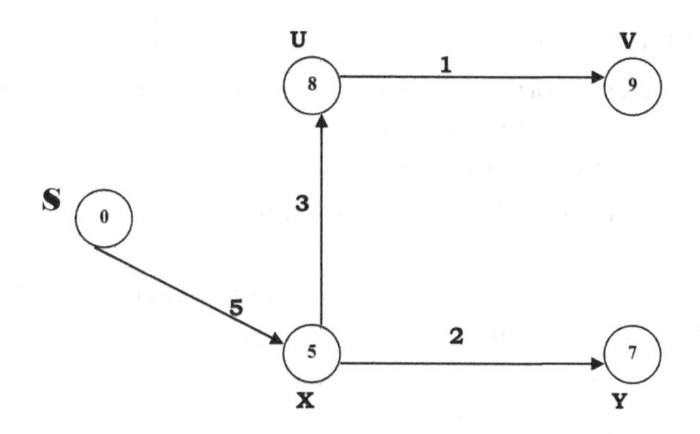

6.4 ALL PAIR SHORTEST PATH

Given a directed, connected weighted graph G(V, E), for each edge $\langle u, v \rangle \in E$, a weight w(u, v) is associated with the edge. The all pairs of shortest paths problem (APSP) is to find the shortest path from u to v for every pair of vertices u and v in V.

- The representation of G

The input is an $n \times n$ matrix W = (wij).

$$w(i,j) = \begin{cases} 0 & \text{if } i = j \\ \text{the weight of the directed edge} \langle i,j \rangle & \text{if } i \neq j \text{ and } \langle i,j \rangle \in E \\ \infty & \text{if } i \neq j \text{ and } \langle i,j \rangle \notin E \end{cases}$$

The all-pairs-shortest-path problem is generalization of the single-source-shortest-path problem, so we can use Floyd's algorithm or Dijkstra's algorithm (varying the source node over all nodes).

- Floyd's algorithm is O(N^3)
- Dijkstra's algorithm with an adjacency matrix is O(N^2), so varying over N source nodes is O(N^3)
- Dijkstra's algorithm with adjacency lists is O(E log N), so varying over N source nodes is O(N E log N)

For large sparse graphs, Dijkstra's algorithm is preferable.

6.4.1 Floyd–Warshall's Algorithm

Floyd–Warshall's algorithm is based upon the observation that a path linking any two vertices u and v may have zero or more intermediate vertices. The algorithm begins by disallowing all intermediate vertices. In this case, the partial solution is simply the initial weights of the graph or infinity if there is no edge.

The algorithm proceeds by allowing an additional intermediate vertex at each step. For each introduction of a new intermediate vertex x, the shortest path between any pair of vertices u and v, x, u, v ∈ V, is the minimum of the previous best estimate of δ(u, v), or the combination of the paths from u → x and x → v.

The Floyd–Warshall algorithm compares all possible paths through the graph between each pair of vertices. It is able to do this with only $\Theta\left(|V|^3\right)$ comparisons in a graph. This is remarkable considering that there may be up to $\Omega\left(|V|^2\right)$ edges in the graph, and every combination of edges is tested. It does so by incrementally improving an estimate on the shortest path between two vertices until the estimate is optimal.

$$\delta(u,v) \leftarrow \min\left(\delta(u,v), \delta(u,x)+\delta(x,v)\right)$$

Let the directed graph be represented by a weighted matrix W.

Floyd–Warshall (W)

```
1.    n ←  rows [W]
2.    D(0) ←  W
3.   for  k ←  1 to  n
4.         do for i ←  1 to  n
5.                     do for j ←  1 to  n
6.                   do  dij(k) ←  MIN (dij(k1), dik(k-1) +
                            dkj(k-1))
7.      return D(n)
```

The time complexity of the algorithm above is O(n3).

Example:

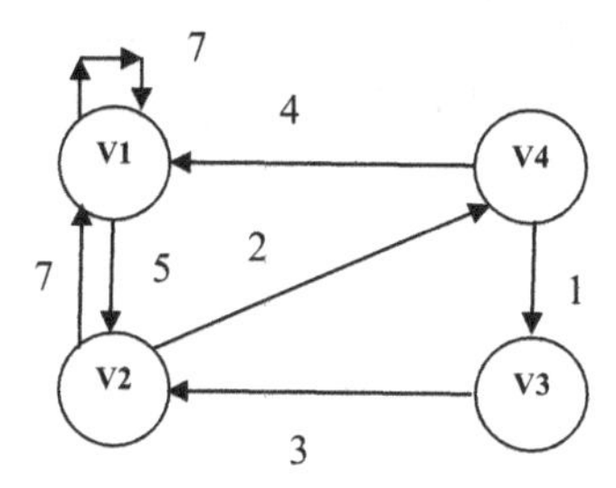

The path matrix is

$$D^0 = \begin{matrix} 7 & 5 & \infty & \infty \\ 7 & \infty & \infty & 2 \\ \infty & 3 & \infty & \infty \\ 4 & \infty & 1 & \infty \end{matrix}$$

According to the algorithm k = 4, i = 4 and j = 4.

So the total number of repetition will be $O(4^3) = 64$.

We have to compute D1, D2, D3, and D4.

For k =1 & I = 1,

J = 1: $D^1[1][1] = \min(D^0[1][1], D^0[1][1] + D^0[1][1]) = (7, 7 + 7) = 7$

J = 2: $D^1[1][2] = \min(D^0[1][2], D^0[1][1] + D^0[1][2]) = (5, 7 + 5) = 5$

J = 3: $D^1[1][3] = \min(D^0[1][3], D^0[1][1] + D^0[1][3]) = (\infty, 7 + \infty) = \infty$

J = 4: $D^1[1][4] = \min(D^0[1][4], D^0[1][1] + D^0[1][4]) = (\infty, 7 + \infty) = \infty$

For k =1 & I = 2,

J = 1: $D^1[2][1] = \min(D^0[2][1], D^0[2][1] + D^0[1][1]) = (7, 7 + 7) = 7$

J = 2: $D^1[2][2] = \min(D^0[2][2], D^0[2][1] + D^0[1][2]) = (\infty, 7 + 5) = \mathbf{12}$

J = 3: $D^1[2][3] = \min(D^0[2][3], D^0[2][1] + D^0[1][3]) = (\infty, 7 + \infty) = \infty$

J = 4: $D^1[2][4] = \min(D^0[2][4], D^0[2][1] + D^0[1][4]) = (2, 7 + \infty) = 2$

For k =1 & I = 3,

J = 1: $D^1[3][1] = \min(D^0[3][1], D^0[3][1] + D^0[1][1]) = (\infty, \infty + 7) = \infty$

J = 2: $D^1[3][2] = \min(D^0[3][2], D^0[3][1] + D^0[1][2]) = (3, \infty + 5) = \mathbf{3}$

J = 3: $D^1[3][3] = \min(D^0[3][3], D^0[3][1] + D^0[1][3]) = (\infty, \infty + \infty) = \infty$

J = 4: $D^1[3][4] = \min(D^0[3][4], D^0[3][1] + D^0[1][4]) = (\infty, \infty + \infty) = \infty$

For k=1 & I = 4,
J = 1: $D^1[4][1] = \min(D^0[4][1], D^0[4][1] + D^0[1][1]) = (4, 4+7) = 4$
J = 2: $D^1[4][2] = \min(D^0[4][2], D^0[4][1] + D^0[1][2]) = (\infty, 4+5) = \mathbf{9}$
J = 3: $D^1[4][3] = \min(D^0[4][3], D^0[4][1] + D^0[1][3]) = (1, 4+\infty) = \mathbf{1}$
J = 4: $D^1[4][4] = \min(D^0[4][4], D^0[4][1] + D^0[1][4]) = (\infty, 4+\infty) = \infty$

$$D^1 = \begin{matrix} 7 & 5 & \infty & \infty \\ 7 & 12 & \infty & 2 \\ \infty & 3 & \infty & \infty \\ 4 & 9 & 1 & \infty \end{matrix}$$

For k=2 & I = 1,
J = 1: $D^2[1][1] = \min(D^1[1][1], D^1[1][2] + D^1[2][1]) = (7, 5+7) = 7$
J = 2: $D^2[1][2] = \min(D^1[1][2], D^1[1][2] + D^1[2][2]) = (5, 5+12) = 5$
J = 3: $D^2[1][3] = \min(D^1[1][3], D^1[1][2] + D^1[2][3]) = (\infty, 5+\infty) = \infty$
J = 4: $D^2[1][4] = \min(D^1[1][4], D^1[1][2] + D^1[2][4]) = (\infty, 5+2) = \mathbf{7}$

For k=2 & I = 2,
J = 1: $D^2[2][1] = \min(D^1[2][1], D^1[2][2] + D^1[2][1]) = (7, 12+7) = 7$
J = 2: $D^2[2][2] = \min(D^1[2][2], D^1[2][2] + D^1[2][2]) = (12, 12+12) = 12$
J = 3: $D^2[2][3] = \min(D^1[2][3], D^1[2][2] + D^1[2][3]) = (\infty, 12+\infty) = \infty$
J = 4: $D^2[2][4] = \min(D^1[2][4], D^1[2][2] + D^1[2][4]) = (2, 12+2) = 2$

For k = 2 & I = 3,
J = 1: $D^2[3][1] = \min(D^1[3][1], D^1[3][2] + D^1[2][1]) = (\infty, 3+7) = \mathbf{10}$
J = 2: $D^2[3][2] = \min(D^1[3][2], D^1[3][2] + D^1[2][2]) = (3, 3+12) = 3$
J = 3: $D^2[3][3] = \min(D^1[3][3], D^1[3][2] + D^1[2][3]) = (\infty, 3+\infty) = \infty$
J = 4: $D^2[3][4] = \min(D^1[3][4], D^1[3][2] + D^1[2][4]) = (\infty, 3+2) = \mathbf{5}$

For k=2 & I = 4,
J = 1: $D^2[4][1] = \min(D^1[4][1], D^1[4][2] + D^1[2][1]) = (4, 9+7) = 4$
J = 2: $D^2[4][2] = \min(D^1[4][2], D^1[4][2] + D^1[2][2]) = (9, 9+12) = 9$
J = 3: $D^2[4][3] = \min(D^1[4][3], D^1[4][2] + D^1[2][3]) = (1, 9+\infty) = 1$
J = 4: $D^2[4][4] = \min(D^1[4][4], D^1[4][2] + D^1[2][4]) = (\infty, 9+2) = \mathbf{11}$

$$D^2 = \begin{matrix} 7 & 5 & \infty & 7 \\ 7 & 12 & \infty & 2 \\ 10 & 3 & \infty & 5 \\ 4 & 9 & 1 & 11 \end{matrix}$$

For k =3 & I = 1,

J = 1: $D^3[1][1] = \min(D^2[1][1], D^2[1][3] + D^2[3][1]) = (7, \infty + 10) = 7$
J = 2: $D^3[1][2] = \min(D^2[1][2], D^2[1][3] + D^2[3][2]) = (5, \infty + 3) = 5$
J = 3: $D^3[1][3] = \min(D^2[1][3], D^2[1][3] + D^2[3][3]) = (\infty, \infty + \infty) = \infty$
J = 4: $D^3[1][4] = \min(D^2[1][4], D^2[1][3] + D^2[3][4]) = (7, \infty + 5) = 7$

For k =3 & I = 2,

J = 1: $D^3[2][1] = \min(D^2[2][1], D^2[2][3] + D^2[3][1]) = (7, \infty + 10) = 7$
J = 2: $D^3[2][2] = \min(D^2[2][2], D^2[2][3] + D^2[3][2]) = (12, \infty + 3) = 12$
J = 3: $D^3[2][3] = \min(D^2[2][3], D^2[2][3] + D^2[3][3]) = (\infty, \infty + \infty) = \infty$
J = 4: $D^3[2][4] = \min(D^2[2][4], D^2[2][3] + D^2[3][4]) = (2, \infty + 5) = 2$

For k = 3 & I = 3,

J = 1: $D^3[3][1] = \min(D^2[3][1], D^2[3][3] + D^2[3][1]) = (10, \infty + 10) = 10$
J = 2: $D^3[3][2] = \min(D^2[3][2], D^2[3][3] + D^2[3][2]) = (3, \infty + 3) = 3$
J = 3: $D^3[3][3] = \min(D^2[3][3], D^2[3][3] + D^2[3][3]) = (\infty, \infty + \infty) = \infty$
J = 4: $D^3[3][4] = \min(D^2[3][4], D^2[3][3] + D^2[3][4]) = (5, \infty + 5) = 5$

For k =3 & I = 4,

J = 1: $D^3[4][1] = \min(D^2[4][1], D^2[4][3] + D^2[3][1]) = (4, 1 + 10) = 4$
J = 2: $D^3[4][2] = \min(D^2[4][2], D^2[4][3] + D^2[3][2]) = (9, 1 + 3) = \mathbf{4}$
J = 3: $D^3[4][3] = \min(D^2[4][3], D^2[4][3] + D^2[3][3]) = (1, 1 + \infty) = 1$
J = 4: $D^3[4][4] = \min(D^2[4][4], D^2[4][3] + D^2[3][4]) = (11, 1 + 5) = \mathbf{6}$

$$D^3 = \begin{matrix} 7 & 5 & \infty & 7 \\ 7 & 12 & \infty & 2 \\ 10 & 3 & \infty & 5 \\ 4 & 4 & 1 & 6 \end{matrix}$$

For k =4 & I = 1,

J = 1: $D^4[1][1] = \min(D^3[1][1], D^3[1][4] + D^3[4][1]) = (7, 7 + 4) = 7$
J = 2: $D^4[1][2] = \min(D^3[1][2], D^3[1][4] + D^3[4][2]) = (5, 7 + 4) = 5$
J = 3: $D^4[1][3] = \min(D^3[1][3], D^3[1][4] + D^3[4][3]) = (\infty, 7 + 1) = \mathbf{8}$
J = 4: $D^4[1][4] = \min(D^3[1][4], D^3[1][4] + D^3[4][4]) = (7, 7 + 6) = 7$

For k =4 & I = 2,

J = 1: $D^4[2][1] = \min(D^3[2][1], D^3[2][4] + D^3[4][1]) = (7, 2 + 4) = \mathbf{6}$
J = 2: $D^4[2][2] = \min(D^3[2][2], D^3[2][4] + D^3[4][2]) = (12, 2 + 4) = \mathbf{6}$
J = 3: $D^4[2][3] = \min(D^3[2][3], D^3[2][4] + D^3[4][3]) = (\infty, 2 + 1) = \mathbf{3}$
J = 4: $D^4[2][4] = \min(D^3[2][4], D^3[2][4] + D^3[4][4]) = (2, 2 + 6) = 2$

For k = 4 & I = 3,

J = 1: $D^4[3][1] = \min(D^3[3][1], D^3[3][4] + D^3[4][1]) = (10, 5 + 4) = \mathbf{9}$

J = 2: $D^4[3][2] = \min(D^3[3][2], D^3[3][4] + D^3[4][2]) = (3, 5 + 4) = 3$

J = 3: $D^4[3][3] = \min(D^3[3][3], D^3[3][4] + D^3[4][3]) = (\infty, 5 + 1) = \mathbf{6}$

J = 4: $D^4[3][4] = \min(D^3[3][4], D^3[3][4] + D^3[4][4]) = (5, 5 + 6) = 5$

For k =4 & I = 4,

J = 1: $D^4[4][1] = \min(D^3[4][1], D^3[4][4] + D^3[4][1]) = (4, 6 + 4) = 4$

J = 2: $D^4[4][2] = \min(D^3[4][2], D^3[4][4] + D^3[4][2]) = (4, 6 + 4) = 4$

J = 3: $D^4[4][3] = \min(D^3[4][3], D^3[4][4] + D^3[4][3]) = (1, 6 + 1) = 1$

J = 4: $D^4[4][4] = \min(D^3[4][4], D^3[4][4] + D^3[4][4]) = (6, 6 + 6) = 6$

$$D^3 = \begin{matrix} 7 & 5 & 8 & 7 \\ 6 & 6 & 3 & 2 \\ 9 & 3 & 6 & 5 \\ 4 & 4 & 1 & 6 \end{matrix}$$

From the above matrix we can plot the graph with all pair shortest path.

The path matrix is

$$D^0 = \begin{matrix} 0 & 8 & \infty & 1 \\ \infty & 0 & 1 & \infty \\ 4 & \infty & 0 & \infty \\ \infty & 2 & 9 & 0 \end{matrix}$$

According to the algorithm k = 4, I = 4 and j = 4.

So the total number of repetition will be $O(4^3) = 64$.

We have to compute D1, D2, D3, and D4.

For k =1 & I = 1,

J = 1: $D^1[1][1] = \min(D^0[1][1], D^0[1][1] + D^0[1][1]) = (0, 0 + 0) = 0$

J = 2: $D^1[1][2] = \min(D^0[1][2], D^0[1][1] + D^0[1][2]) = (8, 0 + 8) = 8$

J = 3: $D^1[1][3] = \min(D^0[1][3], D^0[1][1] + D^0[1][3]) = (\infty, 0 + \infty) = \infty$

J = 4: $D^1[1][4] = \min(D^0[1][4], D^0[1][1] + D^0[1][4]) = (1, 0 + \infty) = 1$

For k =1 & I = 2,

J = 1: $D^1[2][1] = \min(D^0[2][1], D^0[2][1] + D^0[1][1]) = (\infty, \infty + 0) = \infty$

J = 2: $D^1[2][2] = \min(D^0[2][2], D^0[2][1] + D^0[1][2]) = (0, \infty + 8) = 0$

J = 3: $D^1[2][3] = \min(D^0[2][3], D^0[2][1] + D^0[1][3]) = (1, \infty + \infty) = 1$

J = 4: $D^1[2][4] = \min(D^0[2][4], D^0[2][1] + D^0[1][4]) = (\infty, \infty + 1) = \infty$

For k =1 & I = 3,

J = 1: $D^1[3][1] = \min(D^0[3][1], D^0[3][1] + D^0[1][1]) = (4, 4 + 0) = 4$

J = 2: $D^1[3][2] = \min(D^0[3][2], D^0[3][1] + D^0[1][2]) = (\infty, 4 + 8) = \mathbf{12}$

J = 3: $D^1[3][3] = \min(D^0[3][3], D^0[3][1] + D^0[1][3]) = (0, 4 + \infty) = 0$

J = 4: $D^1[3][4] = \min(D^0[3][4], D^0[3][1] + D^0[1][4]) = (\infty, 4 + 1) = \mathbf{5}$

For k =1 & I = 4,

J = 1: $D^1[4][1] = \min(D^0[4][1], D^0[4][1] + D^0[1][1]) = (\infty, \infty + 0) = \infty$

J = 2: $D^1[4][2] = \min(D^0[4][2], D^0[4][1] + D^0[1][2]) = (2, \infty + 8) = \mathbf{2}$

J = 3: $D^1[4][3] = \min(D^0[4][3], D^0[4][1] + D^0[1][3]) = (9, \infty + \infty) = \mathbf{9}$

J = 4: $D^1[4][4] = \min(D^0[4][4], D^0[4][1] + D^0[1][4]) = (0, \infty + 1) = 0$

$$D^1 = \begin{matrix} 0 & 8 & \infty & 1 \\ \infty & 0 & 1 & \infty \\ 4 & 12 & 0 & 5 \\ \infty & 2 & 9 & 0 \end{matrix}$$

For k =2 & I = 1,

J = 1: $D^2[1][1] = \min(D^1[1][1], D^1[1][2] + D^1[2][1]) = (0, 8 + \infty) = 0$

J = 2: $D^2[1][2] = \min(D^1[1][2], D^1[1][2] + D^1[2][2]) = (8, 8 + 0) = 8$

J = 3: $D^2[1][3] = \min(D^1[1][3], D^1[1][2] + D^1[2][3]) = (\infty, 8 + 1) = \mathbf{9}$

J = 4: $D^2[1][4] = \min(D^1[1][4], D^1[1][2] + D^1[2][4]) = (1, 8 + \infty) = 1$

For k =2 & I = 2,

J = 1: $D^2[2][1] = \min(D^1[2][1], D^1[2][2] + D^1[2][1]) = (\infty, 0 + \infty) = \infty$

J = 2: $D^2[2][2] = \min(D^1[2][2], D^1[2][2] + D^1[2][2]) = (0, 0 + 0) = 0$

J = 3: $D^2[2][3] = \min(D^1[2][3], D^1[2][2] + D^1[2][3]) = (1, 0 + 1) = 1$

J = 4: $D^2[2][4] = \min(D^1[2][4], D^1[2][2] + D^1[2][4]) = (\infty, 0 + \infty) = \infty$

For k = 2 & I = 3,

J = 1: $D^2[3][1] = \min(D^1[3][1], D^1[3][2] + D^1[2][1]) = (4, 12 + \infty) = 4$

J = 2: $D^2[3][2] = \min(D^1[3][2], D^1[3][2] + D^1[2][2]) = (12, 12 + 0) = 12$

J = 3: $D^2[3][3] = \min(D^1[3][3], D^1[3][2] + D^1[2][3]) = (0, 12 + 1) = 0$

J = 4: $D^2[3][4] = \min(D^1[3][4], D^1[3][2] + D^1[2][4]) = (5, 12 + \infty) = 5$

For k =2 & I = 4,

J = 1: $D^2[4][1] = \min(D^1[4][1], D^1[4][2] + D^1[2][1]) = (\infty, 2 + \infty) = \infty$

J = 2: $D^2[4][2] = \min(D^1[4][2], D^1[4][2] + D^1[2][2]) = (2, 2 + 0) = 2$

J = 3: $D^2[4][3] = \min(D^1[4][3], D^1[4][2] + D^1[2][3]) = (9, 2 + 1) = \mathbf{3}$

J = 4: $D^2[4][4] = \min(D^1[4][4], D^1[4][2] + D^1[2][4]) = (0, 2 + \infty) = 0$

$$D^2 = \begin{matrix} 0 & 8 & 9 & 1 \\ \infty & 0 & 1 & \infty \\ 4 & 12 & 0 & 5 \\ \infty & 2 & 3 & 0 \end{matrix}$$

For k =3 & I = 1,

J = 1: $D^3[1][1] = \min(D^2[1][1], D^2[1][3] + D^2[3][1]) = (0, 9 + 4) = 0$
J = 2: $D^3[1][2] = \min(D^2[1][2], D^2[1][3] + D^2[3][2]) = (8, 9 + 12) = 8$
J = 3: $D^3[1][3] = \min(D^2[1][3], D^2[1][3] + D^2[3][3]) = (9, 9 + 0) = 9$
J = 4: $D^3[1][4] = \min(D^2[1][4], D^2[1][3] + D^2[3][4]) = (1, 9 + 5) = 1$

For k =3 & I = 2,

J = 1: $D^3[2][1] = \min(D^2[2][1], D^2[2][3] + D^2[3][1]) = (\infty, 1 + 4) = \mathbf{5}$
J = 2: $D^3[2][2] = \min(D^2[2][2], D^2[2][3] + D^2[3][2]) = (0, 1 + 12) = 0$
J = 3: $D^3[2][3] = \min(D^2[2][3], D^2[2][3] + D^2[3][3]) = (1, 1 + 0) = 1$
J = 4: $D^3[2][4] = \min(D^2[2][4], D^2[2][3] + D^2[3][4]) = (\infty, 1 + 5) = \mathbf{6}$

For k = 3 & I = 3,

J = 1: $D^3[3][1] = \min(D^2[3][1], D^2[3][3] + D^2[3][1]) = (4, 0 + 4) = 4$
J = 2: $D^3[3][2] = \min(D^2[3][2], D^2[3][3] + D^2[3][2]) = (12, 0 + 12) = 12$
J = 3: $D^3[3][3] = \min(D^2[3][3], D^2[3][3] + D^2[3][3]) = (0, 0 + 0) = 0$
J = 4: $D^3[3][4] = \min(D^2[3][4], D^2[3][3] + D^2[3][4]) = (5, 0 + 5) = 5$

For k =3 & I = 4,

J = 1: $D^3[4][1] = \min(D^2[4][1], D^2[4][3] + D^2[3][1]) = (\infty, 3 + 4) = \mathbf{7}$
J = 2: $D^3[4][2] = \min(D^2[4][2], D^2[4][3] + D^2[3][2]) = (2, 3 + 12) = 2$
J = 3: $D^3[4][3] = \min(D^2[4][3], D^2[4][3] + D^2[3][3]) = (3, 3 + 0) = 3$
J = 4: $D^3[4][4] = \min(D^2[4][4], D^2[4][3] + D^2[3][4]) = (0, 3 + 5) = 0$

$$D^3 = \begin{matrix} 0 & 8 & 9 & 1 \\ 5 & 0 & 1 & 6 \\ 4 & 12 & 0 & 5 \\ 7 & 2 & 3 & 0 \end{matrix}$$

For k =4 & I = 1,

J = 1: $D^4[1][1] = \min(D^3[1][1], D^3[1][4] + D^3[4][1]) = (0, 1 + 7) = 0$
J = 2: $D^4[1][2] = \min(D^3[1][2], D^3[1][4] + D^3[4][2]) = (8, 1 + 2) = \mathbf{3}$
J = 3: $D^4[1][3] = \min(D^3[1][3], D^3[1][4] + D^3[4][3]) = (9, 1 + 3) = \mathbf{4}$
J = 4: $D^4[1][4] = \min(D^3[1][4], D^3[1][4] + D^3[4][4]) = (1, 1 + 0) = 1$

For k =4 & I = 2,

J = 1: $D^4[2][1] = \min(D^3[2][1], D^3[2][4] + D^3[4][1]) = (5, 6 + 7) = 5$
J = 2: $D^4[2][2] = \min(D^3[2][2], D^3[2][4] + D^3[4][2]) = (0, 6 + 2) = 0$
J = 3: $D^4[2][3] = \min(D^3[2][3], D^3[2][4] + D^3[4][3]) = (1, 6 + 3) = 1$
J = 4: $D^4[2][4] = \min(D^3[2][4], D^3[2][4] + D^3[4][4]) = (6, 6 + 0) = 6$

For k = 4 & I = 3,

J = 1: $D^4[3][1] = \min(D^3[3][1], D^3[3][4] + D^3[4][1]) = (4, 5 + 7) = 4$
J = 2: $D^4[3][2] = \min(D^3[3][2], D^3[3][4] + D^3[4][2]) = (12, 5 + 2) = 7$
J = 3: $D^4[3][3] = \min(D^3[3][3], D^3[3][4] + D^3[4][3]) = (0, 5 + 3) = 0$
J = 4: $D^4[3][4] = \min(D^3[3][4], D^3[3][4] + D^3[4][4]) = (5, 5 + 0) = 5$

For k =4 & I = 4,

J = 1: $D^4[4][1] = \min(D^3[4][1], D^3[4][4] + D^3[4][1]) = (7, 0 + 7) = 7$
J = 2: $D^4[4][2] = \min(D^3[4][2], D^3[4][4] + D^3[4][2]) = (2, 0 + 2) = 2$
J = 3: $D^4[4][3] = \min(D^3[4][3], D^3[4][4] + D^3[4][3]) = (3, 0 + 3) = 3$
J = 4: $D^4[4][4] = \min(D^3[4][4], D^3[4][4] + D^3[4][4]) = (0, 0 + 0) = 0$

The all pair shortest path matrix is

$$D^4 = \begin{matrix} 0 & 3 & 4 & 1 \\ 5 & 0 & 1 & 6 \\ 4 & 7 & 0 & 5 \\ 7 & 2 & 3 & 0 \end{matrix}$$

6.5 QUESTIONS

6.5.1 Short Questions

1. What are BFS and DFS?
2. Explain the relaxation technique.
3. How many types of graph traversal and write the names?
4. Describe where BFS is used.
5. What is spanning tree?
6. Explain minimum spanning tree.
7. Explain Kruskal's algorithm.

8. Define Prim's algorithm.
9. Define negative weighted edges.
10. What is a relaxation technique?

6.5.2 Long Questions

1. Write the Bellman–Ford algorithm.
2. Write Dijkstara's algorithm with an example.
3. Write Floyd–Warshall algorithm and explain time complexity.
4. Explain Dijkstra's algorithm.
5. Explain relaxation technique with an algorithm.
6. Write the BFS algorithm with an example.
7. How does Kruskal's algorithm know when the addition of an edge will generate a cycle?
8. Explain the Bellman–Ford algorithm with proper example.
9. Describe the Floyd–Warshall algorithm with an example.
10. Solve this problem using Prim's algorithm.

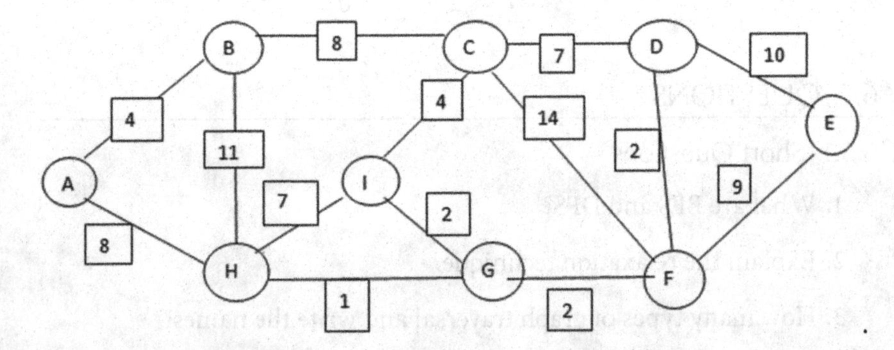

CHAPTER 7

Approximation Algorithms

7.1 HAMILTONIAN CYCLE

A Hamiltonian cycle of a directed graph G = (V, E) is a simple cycle that contains each vertex in V. A graph that contains a Hamiltonian cycle is said to be Hamiltonian; otherwise, it is non-Hamiltonian.

Example:

The Black represented edges create the Hamiltonian cycle.

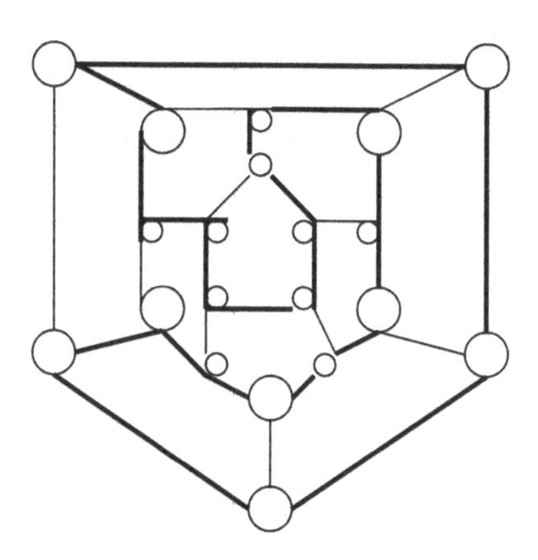

DOI: 10.1201/9781003093886-7

Example:

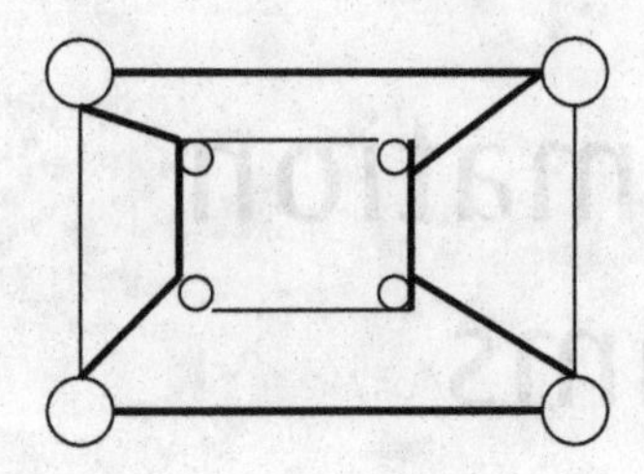

7.2 APPROXIMATION ALGORITHMS

7.2.1 Traveling Salesman Problem

Given a graph G = (V, E) non-negative edge weight C(e) and an integer C, there is a Hamiltonian cycle whose total cost is at most C.

In other words, we can define TSP as a salesman who is required to visit a number of cities during the trip, given the distance between the cities, in what order should he travel to visit every city precisely once and returned home, with the minimum mileage traveled.

7.2.1.1 Special Case of TSP

Here first of all we define optimal TSP as: "Given a complete weighted graph find a cycle of the minimum cost that visits each vertex".

The opt-TSP is NP-Hard

Special case:

Edge weights satisfy the triangle inequality as

$$W(a,b)+W(b,c)\geq W(a,c)$$

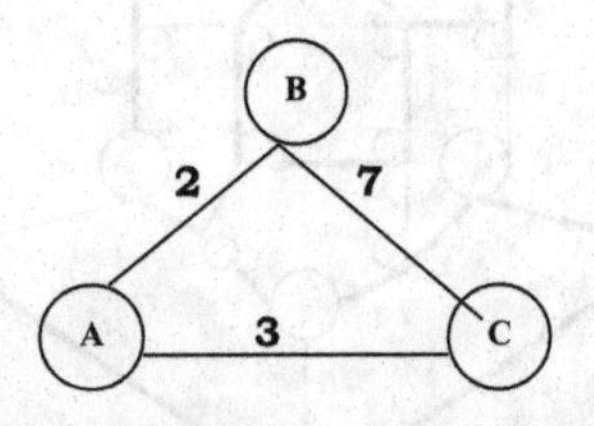

7.2.1.1.1 TSP with Triangle Inequality

Developing an approximate algorithm for TSP is impossible unless P = NP.

For the below algorithm,

Input: Weighted complete graph G satisfying triangle inequality

Output: A TSP tour T for G

ALGORITHM TSP(G)

```
1. m ← A minimum spanning tree of G
2. p ← An Euler tour traversal i.e./Preorder traversal
3. T ← Empty
4. for each vertex v in P (in preorder traversal)
5. if this is v's first appearance in P
6. then T. insert last(v)
7. T.insert last(S)
8. Return T
```

Now connect the edges according to the preorder traversal (Figures 7.1–7.4).

PREORDER: A	B	C	H	D	E	F	G	
Hamiltonian Cycle concept: A	B	C	H	D	E	F	G	A

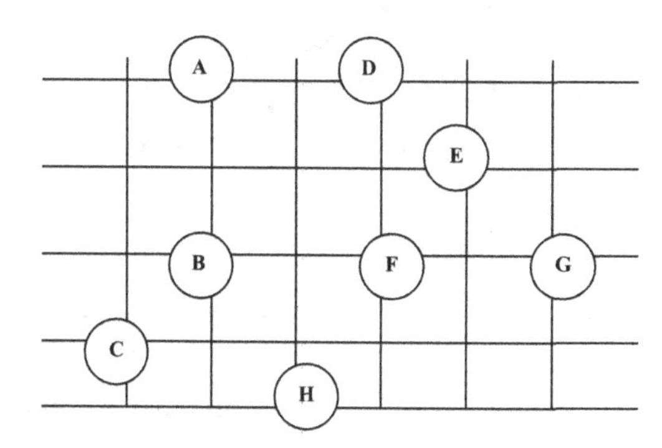

FIGURE 7.1 Given set of data.

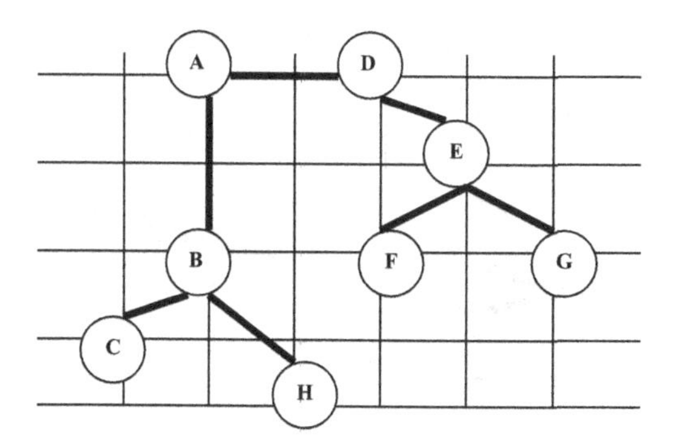

FIGURE 7.2 Minimum Spanning Tree formed according to the distance between the vertices starting from A.

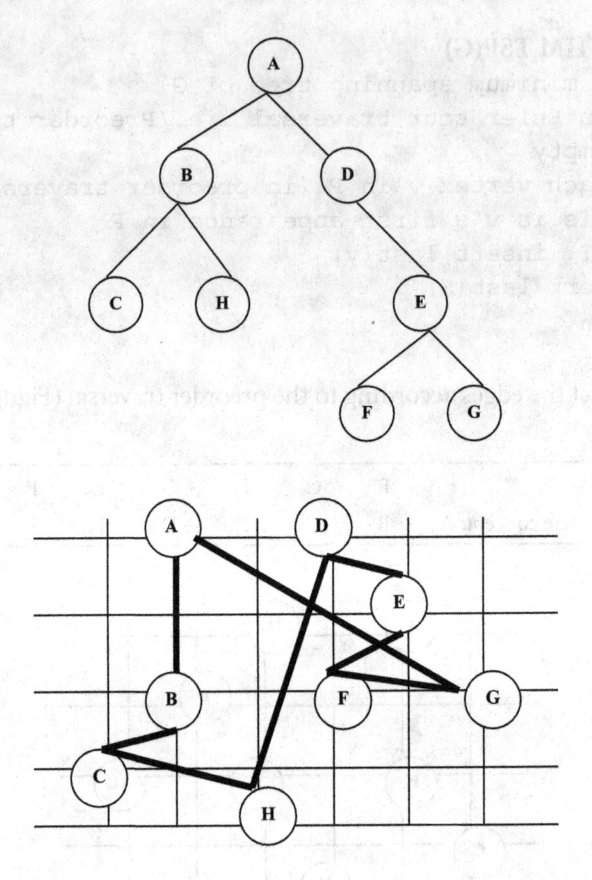

FIGURE 7.3 Vertices connected according to preorder traversal, but it is not optimal so implement the triangle inequality or Hamiltonian cycle.

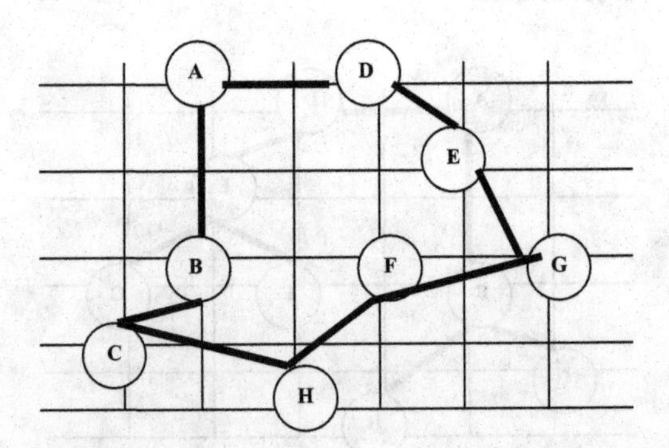

FIGURE 7.4 TSP solution.

By implementing these the edge (H, G) and (G, A) will too expensive so remove them and connect to (H, F), (F, G), (G, E), (E, D) and (D, A).

Now the optimal TSP solution is

7.3 BACKTRACKING

The main concept for the backtracking problem depends on the binary choice, which means yes or no. Whenever the backtracking has choice no that means the algorithm has encountered a dead end and it backtracks one step and tries a different path for choice yes. The backtracking resembles a DFS tree in a directed graph where the graph is either a tree or at least it does not have any cycles.

Knapsack Problem

ALGORITHM BackKnapsack(I, W)

I: Items W: KnapSack Capacity

```
1. SET B<- 0
2. FOR I <- 1 TO N
3. IF (wi <= W) then
4. B<= max (B, vi + BackKnapsack(I, W - wi))
5. Return B
```

Example:

Let consider the items

I = <I1, I2, I3>

w = <2, 3, 4>

V = <3, 4, 5>

And the Knapsack capacity is W = 5.

Solution:

To solve the knapsack problem using backtracking, we have to plot the tree or graph. The structure of the node will be

;0,0

where the left-hand side of the semicolon represents the weight chosen and the first entry to the right-hand side of the semicolon corresponds to the total value and the second entry corresponds to the total weight concerning the left side of the semicolon.

Initially, our solution will start from empty. Later on, at every mode down to the children, we select the type of the item to be added to the knapsack.

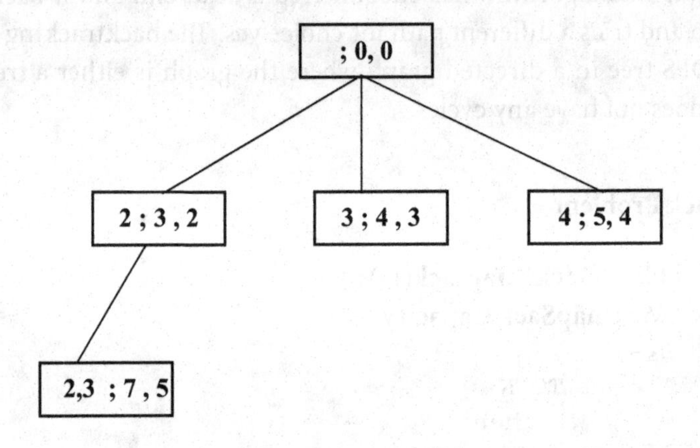

7.3.1 Hamiltonian Circuit Problem

Given a graph G = (V, E), we have to find the Hamiltonian cycle. Using the backtracking approach, we will start our search from an arbitrary vertex, which will be treated as our root. The next adjacent vertex is selected based on the alphabetical or numerical order. If at any stage, the arbitrary vertex will form a cycle with the root, then we can say that as a dead end. In this case, we backtrack one step, and again the search will begin by selecting another vertex. It should be noted that after backtracking the element from the partial solution, it must be removed.

Example:

Consider the graph

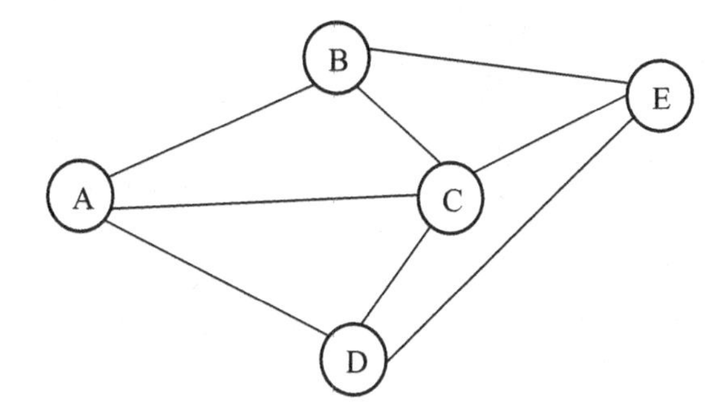

Solution:

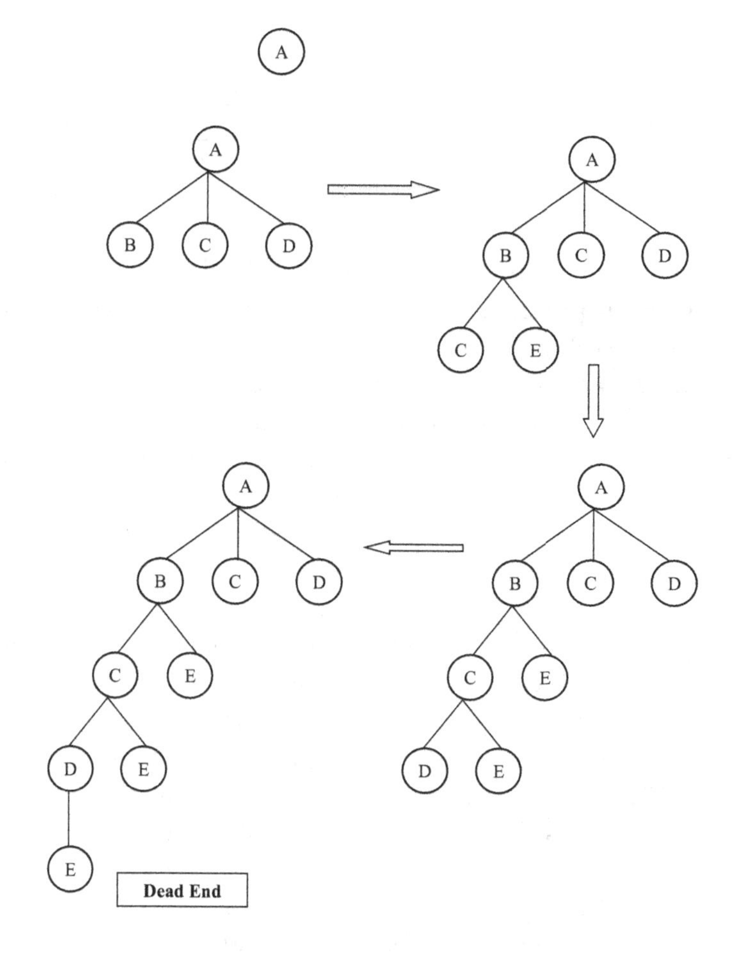

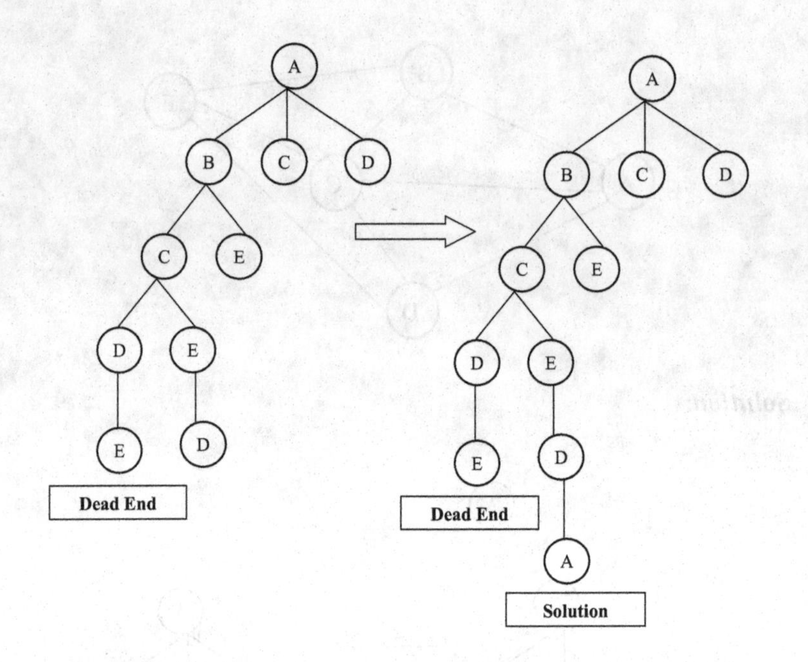

7.4 N-QUEEN PROBLEM/8 – QUEEN PROBLEM

The **eight queens puzzle** is the problem of placing eight chess queens on an 8 × 8 chessboard so that no two queens attack each other. Thus, a solution requires that no two queens share the same row, column, or diagonal. The eight queens puzzle is an example of the more general n-queen problem of placing n queens on an n × n chessboard, where solutions exist for all natural numbers n except for n = 2 and n = 3.

The problem can be quite computationally expensive as there are 4,426,165,368 (i.e., 64 choose 8) possible arrangements of eight queens on an 8×8 board, but only 92 solutions. It is possible to use shortcuts that reduce computational requirements or rules of thumb that avoid brute-force computational techniques. For example, just by applying a simple rule that constrains each queen to a single column (or row), though still considered brute force, it is possible to reduce the number of possibilities to just 16,777,216 (i.e., 88) possible combinations. Generating permutations further reduces the possibilities to just 40,320 (i.e., 8!), which are then checked for diagonal attacks.

These brute-force algorithms are computationally manageable for n = 8 but would be intractable for problems of n ≥ 20, as 20! = 2.433 * 1018. Advancements for this and other toy problems are the development and application of heuristics (rules of thumb) that yield solutions to the n queens puzzle at a small fraction of the computational requirements.

This heuristic solves N queens for any $N \geq 4$. It forms the list of numbers for vertical positions (rows) of queens with horizontal position (column) simply increasing. N is 8 for eight queens puzzle.

1. If the remainder from dividing N by 6 is not 2 or 3, then the list is simply all even numbers followed by all odd numbers $\leq N$.
2. Otherwise, write separate lists of even and odd numbers (i.e., 2, 4, 6, 8 – 1, 3, 5 and 7).
3. If the remainder is 2, swap 1 and 3 in the odd list and move 5 to the end (i.e., **3, 1**, 7 and **5**).
4. If the remainder is 3, move 2 to the end of the even list and 1, 3 to the end of the odd list (i.e., 4, 6, 8, **2** – 5, 7, 9, **1** and **3**).
5. Append odd list to the even list and place queens in the rows given by these numbers, from left to right (i.e. a2, b4, c6, d8, e3, f1, g7 and h5).

For $N = 8$, this results in the solution shown above. A few more examples follow:

- 14 queens (remainder 2): 2, 4, 6, 8, 10, 12, 14, 3, 1, 7, 9, 11, 13 and 5.
- 15 queens (remainder 3): 4, 6, 8, 10, 12, 14, 2, 5, 7, 9, 11, 13, 15, 1 and 3.
- 20 queens (remainder 2): 2, 4, 6, 8, 10, 12, 14, 16, 18, 20, 3, 1, 7, 9, 11, 13, 15, 17, 19 and 5.

Solutions

The eight queens puzzle has 92 **distinct** solutions. If solutions that differ only by symmetry operations (rotations and reflections) of the board are counted as one, the puzzle has 12 **unique** (or **fundamental**) solutions.

A fundamental solution usually has eight variants (including its original form) obtained by rotating 90°, 180° or 270° and then reflecting each of the four rotational variants in a mirror in a fixed position. However, should a solution be equivalent to its own 90° rotation (as happens to one solution with five queens on a 5×5 board) that fundamental solution will have only two variants (itself and its reflection). Should a solution be equivalent to its own 180° rotation (but not to its 90° rotation) it will have four variants (itself, its reflection, its 90° rotation and the reflection of that). A solution can't be equivalent to its own reflection (except at $n = 1$) because that would require two queens to be facing each other. Of the 12 fundamental solutions to the problem with eight queens on an 8×8 board, exactly one is equal to its own 180° rotation, and none are equal to their 90° rotation, thus the number of distinct solutions is $11 * 8 + 1 * 4 = 92$ (where the 8 is derived from four 90° rotational positions and their reflections, and the 4 is derived from two 180° rotational positions and their reflections).

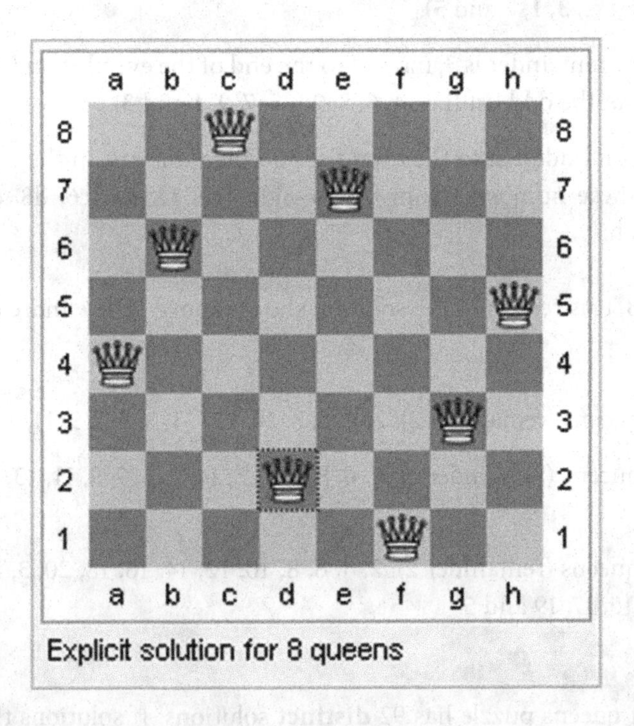

Explicit solution for 8 queens

The examples above can be obtained with the following formulas. Let (i, j) be the square in column i and row j on the $n \times n$ chessboard, k an integer.

1. If n is even and $n \neq 6k + 2$, then place queens at (i, 2i) and (n/2 + i, 2i − 1) for i = 1, 2, …, n/2.

2. If n is even and $n \neq 6k$, then place queens at $(i, 1 + (2i + n/2 - 3 \pmod n))$ and $(n + 1 - i, n - (2i + n/2 - 3 \pmod n))$ for $i = 1, 2, \ldots, n/2$.
3. If n is odd, then use one of the patterns above for $(n - 1)$ and add a queen at (n, n).

7.5 BACKTRACKING ALGORITHM

The idea is to place queens one by one in different columns, starting from the leftmost column. When we place a queen in a column, we check for clashes with already placed queens. In the current column, if we find a row for which there is no clash, we mark this row and column as part of the solution. If we do not find such a row due to clashes, then we backtrack and return false.

1. Start in the leftmost column
2. If all queens are placed
   return true
3. Try all rows in the current column. Do following for every tried row.
   a. If the queen can be placed safely in this row then mark this [row, column] as part of the solution and recursively check if placing queen here leads to a solution.
   b. If placing queen in [row, column] leads to a solution then return true.
   c. If placing queen doesn't lead to a solution then unmark this [row, column] (Backtrack) and go to step (a) to try other rows.
4. If all rows have been tried and nothing worked, return false to trigger backtracking.

7.6 BRANCH AND BOUND

Branch and bound (**BB** or **B&B**) is a general algorithm for finding optimal solutions to various optimization problems, especially in discrete and combinatorial optimization. A branch-and-bound algorithm consists of a systematic enumeration of all candidate solutions, where large subsets of fruitless candidates are discarded *en masse*, by using upper and lower estimated bounds of the quantity being optimized.

7.6.1 Knapsack Problem

In branch bound, the problems are being solved by using the tree data structure. The structure of a node divided into three parts:

1. The first part indicates the total weight of the item.
2. The second part indicates the value of the current item.
3. The third part indicates the upper bound for the node.

W	V
Ub	

The upper bound for a node can be computed as

$$Ub = v + (W - w)(v_{i+1}/w_{i+1})$$

Example:

Given three items as

Item:	11	12	13
Weights:	1	2	3
Values:	2	3	4

With knapsack weight W = 3

Solution:

Items	w_i	v_i	V_i/w_i
I1	1	2	2
I2	2	3	1.5
I3	3	4	1.3

Now start with the root node with weight = 0 and value = 0 and upper bound = 6

$$Ub = 0 + (3 - 0) * 2 = 6 \quad [v = 0, w = 0, W = 3, v1/w1 = 2]$$

w = 0	v = 0
Ub = 6	

Next, we include item 1 which is indicated by the left branch and exclude 1 which is indicated by the right branch.

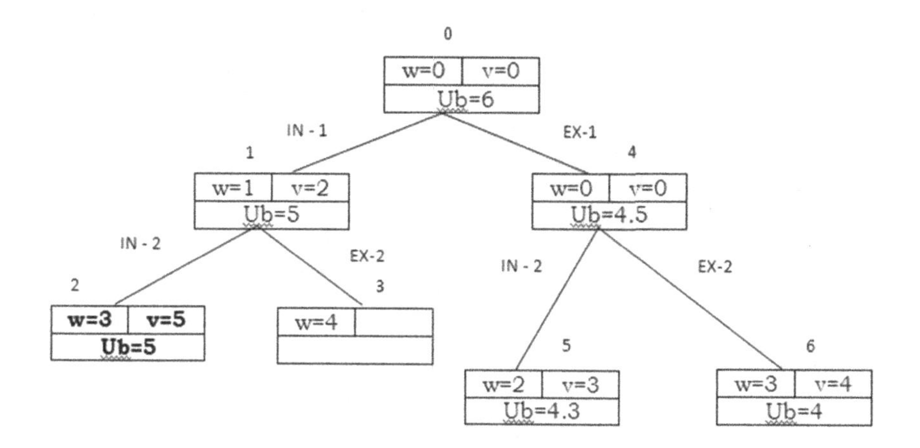

Example:

Consider the following BIP problem:

$$\text{Max } Z = 9x_1 + 5x_2 + 6x_3 + 4x_4$$

$$\text{s.t. } 6x_1 + 3x_2 + 5x_3 + 2x_4 \leq 10$$

$$x_3 + x_4 \leq 1$$

$$-x_1 + x_3 \leq 0$$

$$-x_2 + x_4 \leq 0$$

$$x_i \text{ are binary}$$

The optimal solution of the LP relaxation

$$\text{Max } Z = 9x_1 + 5x_2 + 6x_3 + 4x_4$$

$$\text{s.t. } 6x_1 + 3x_2 + 5x_3 + 2x_4 \le 10$$

$$x_3 + x_4 \le 1$$

$$-x_1 + x_3 \le 0$$

$$-x_2 + x_4 \le 0$$

$$x_i \le 1 \quad \text{for } 1 \le i \le 4$$

$$x_i \ge 0$$

has optimal solution at (5/6, 1, 0, 1) with Z = 16.5.

Branch 1: $x_1 = 0$ or $x_1 = 1$

1. $x_1 = 0$. The problem becomes

$$\text{Max } Z = 5x_2 + 4x_4$$

Subject to

$$3x_2 + 2x_4 \le 10$$

$$-x_2 + x_4 \le 0$$

$$x_i \text{ are binary}$$

The optimal solution of the LP relaxation is at (1, 1) with Z = 9. (Current best solution.)

2. $x_1 = 0$. The LP relaxation

$$\text{Max } Z = 9 + 5x_2 + 6x_3 + 4x_4$$

$$\text{s.t. } 3x_2 + 5x_3 + 2x_4 \le 4$$

$$x_3 + x_4 \le 1$$

$$x_3 \le 1$$

$$-x_2 + x_4 \le 0$$

$$x_i \le 1 \quad \text{for } 2 \le i \le 4$$

$$x_i \ge 0$$

has optimal solution at (1, 0.8, 0, 0.8) with Z = 16.2.

Branch: $x_2 = 0$ or $x_2 = 1$.

2.1 $x_2 = 0$. The LP relaxation (in this case we have $x_4 = 0$ as well)

$$\text{Max } Z = 9 + 6x_3$$

$$\text{s.t. } 5x_3 \le 4$$

$$x_3 \le 1$$

$$x_3 \ge 0$$

has optimal solution at (1, 0, 0.8, 0) with Z = 13.8.

Branch: $x_3 = 0$ or $x_3 = 1$.

2.1.1 $x_3 = 0$. The optimal solution is (1, 0, 0, 0) with Z = 9 (not better than the current best solution).

2.1.2 $x_3 = 1$. Not feasible.

2.2 $x_2 = 1$. The optimal solution of the LP relaxation

$$\text{Max } Z = 14 + 6x_3 + 4x_4$$

$$\text{s.t. } 5x_3 + 2x_4 \le 1$$

$$x_3 + x_4 \le 1$$

$$x_3 \le 1$$

$$x_4 \le 1$$

$$x_i \ge 0$$

is at (1, 1, 0, 0.5) with Z = 16.

2.2.1 $x_3 = 0$. The previous optimal solution (1, 1, 0, 0.5) is still feasible therefore still optimal.

2.2.1.1 $x_4 = 0$. (1, 1, 0, 0) is feasible. Z = 14. (It becomes the current best solution.)

2.2.1.2 $x_4 = 1$. (1, 1, 0, 1) is not feasible.

2.2.2 $x_3 = 1$. No feasible solution.

Therefore the current best solution (1, 1, 0, 0) with Z = 14 is the optimal solution.

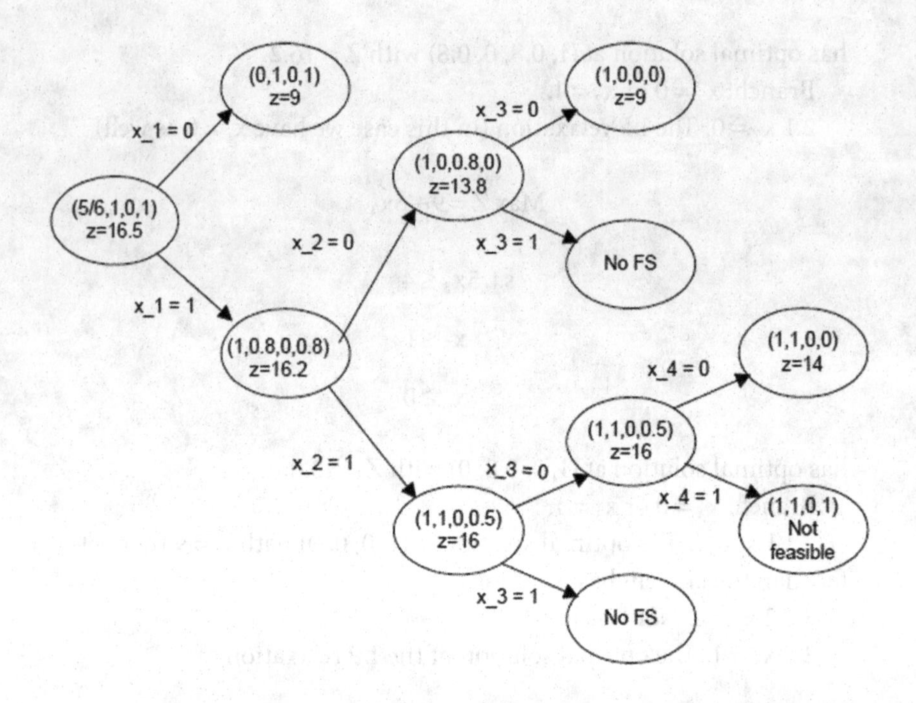

7.7 QUESTIONS

7.7.1 Short Questions

1. What is Hamiltonian cycle?
2. What is branch and bound algorithm?
3. What is backtracking?
4. What is NP completeness?
5. Explain TSP.
6. What is NP hard?
7. What Hamiltonian circuit?
8. Explain branch and bound technique.
9. What is knapsack problem?
10. What is queen problem?

7.7.2 Long Questions

1. Write back the knapsack algorithm with an example.
2. Short notes on
   - Knapsack problem
   - NP completeness
   - NP hard
3. Short notes on
   - Branch and bound
   - Hamiltonian circuit
4. Write backtracking algorithm.
5. Explain Approximation algorithm.
6. Consider the graph.

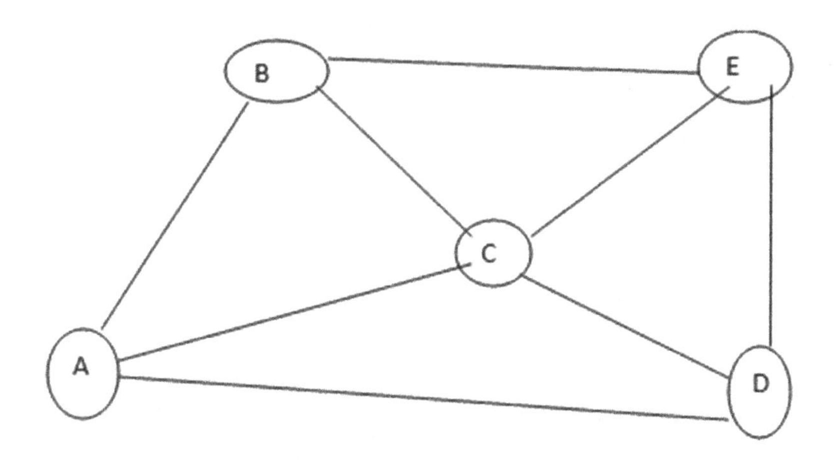

   Using Hamiltonian circuit problem

7. Solve the eight-queen problem.
8. Explain backtracking. Write algorithm with proper example.
9. Write TSP algorithm with proper example.

10. Explain knapsack problem. Solve the given problem
    Item: 11 12 13
    Weights: 1 2 3
    Values: 2 3 4
    With knapsack weight W = 3.

CHAPTER 8

Matrix Operations, Linear Programming, Polynomial and FFT

8.1 MATRICES

A matrix is a rectangular array of m×n elements, where m is the number of rows and n is the number of columns in the matrix. In general, a matrix is represented as **A**.

The elements $\mathbf{a}_{ij}$ means the element of ith row and jth column.

- If a matrix having a single row, then it is called **row matrix**.
- If a matrix having a single column, then it is called **column matrix**.
- If in a matrix the number of row and number of column are the same, then that matrix is called **square matrix**.
- If in a matrix all the row and column elements are 0, then that matrix is known as **additive identity matrix** and represented as **0**.
- An **identity matrix** is denoted as **I**, is a square matrix with 1's in diagonal and 0's elsewhere.

DOI: 10.1201/9781003093886-8

Examples:

Row Matrix $\begin{bmatrix} 5 & 7 & 8 & 12 \end{bmatrix}$

Column Matrix $\begin{pmatrix} 5 \\ 7 \\ 8 \\ 12 \end{pmatrix}$

Square Matrix $\begin{pmatrix} 5 & 6 & 9 \\ 7 & 8 & 2 \\ 8 & 5 & 4 \end{pmatrix}$

Additive Identity Matrix(0) $\begin{pmatrix} 0 & 0 & 0 \\ 0 & 0 & 0 \\ 0 & 0 & 0 \end{pmatrix}$

Identity Matrix $\begin{pmatrix} 1 & 0 & 0 \\ 0 & 1 & 0 \\ 0 & 0 & 1 \end{pmatrix}$

8.1.1 Operations with Matrix

For the matrix, we can implement four operations such as

- Addition
- Subtraction
- Multiplication
- Scalar multiplication

Two matrices are said to be equal if they have the same number of rows and the same number of columns and corresponding elements are equal.

Mathematically, $A = B$ iff $a_{ij} = b_{ij}$ for all i & j.

Addition

$$\begin{pmatrix} 2 & 3 & 4 \\ 5 & 6 & 7 \\ 2 & 2 & 2 \end{pmatrix} + \begin{pmatrix} 1 & 6 & 2 \\ 5 & 1 & 3 \\ 4 & 8 & 7 \end{pmatrix} = \begin{pmatrix} 3 & 9 & 6 \\ 10 & 7 & 10 \\ 6 & 10 & 9 \end{pmatrix}$$

Subtraction

$$\begin{pmatrix} 2 & 3 & 4 \\ 5 & 6 & 7 \\ 2 & 2 & 2 \end{pmatrix} - \begin{pmatrix} 1 & 6 & 2 \\ 5 & 1 & 3 \\ 4 & 8 & 7 \end{pmatrix} = \begin{pmatrix} 1 & -3 & 2 \\ 0 & 5 & 4 \\ -2 & -6 & -5 \end{pmatrix}$$

Multiplication

$$\begin{pmatrix} 2 & 3 & 4 \\ 5 & 6 & 7 \\ 2 & 2 & 2 \end{pmatrix} * \begin{pmatrix} 1 & 6 & 2 \\ 5 & 1 & 3 \\ 4 & 8 & 7 \end{pmatrix} = \begin{pmatrix} 33 & 47 & 41 \\ 63 & 92 & 77 \\ 20 & 30 & 24 \end{pmatrix}$$

Scalar Multiplication

$$5 * \begin{pmatrix} 1 & 6 & 2 \\ 5 & 1 & 3 \\ 4 & 8 & 7 \end{pmatrix} = \begin{pmatrix} 5 & 30 & 10 \\ 25 & 5 & 15 \\ 20 & 40 & 35 \end{pmatrix}$$

Determinant

The determinant of a square matrix A of size mxm is denoted as det(A) is a scalar which is calculated as

1. If $m=1$, $\det(A)=a_{11}$
2. if $m>1$, $\det(A) = \sum_{i=1\ldots m} (-1)^{i+j} \times a_{ij} \times \det(A_{ij})$

   where Aij is a matrix obtained from A by deleting ith row and jth column.

Process to compute the Determinant

For a 1×1 matrix, the determinant is the only item of the matrix.

Example: $|1|=1$

For 2×2 matrix, the calculation is:

$$\begin{vmatrix} a & b \\ c & d \end{vmatrix} = ad - bc$$

For a 3×3 matrix, the calculation is:

$$\begin{vmatrix} a & b & c \\ d & e & f \\ g & h & i \end{vmatrix} = a\begin{vmatrix} e & f \\ h & i \end{vmatrix} - b\begin{vmatrix} d & f \\ g & i \end{vmatrix} + c\begin{vmatrix} d & e \\ g & h \end{vmatrix}$$

$$= aei - afh + bfg - bdi + cdh - ceg$$

Example 1:

Compute the determinant of

$$A = \begin{pmatrix} 2 & 3 & 2 \\ 4 & 5 & 2 \\ 5 & 6 & 2 \end{pmatrix}$$

Solution

$$\det \mathbf{A} = \begin{vmatrix} 2 & 3 & 2 \\ 4 & 5 & 2 \\ 5 & 6 & 2 \end{vmatrix} = 2\cdot5\cdot2 + 3\cdot2\cdot5 + 2\cdot4\cdot6 - 2\cdot5\cdot5 - 2\cdot2\cdot6 - 3\cdot4\cdot2$$

$$= 20 + 30 + 48 - 50 - 24 - 24 = 0$$

Example 2:

Compute the determinant of

$$\mathbf{A} = \begin{pmatrix} 1 & 2 & 4 & 5 \\ 8 & 5 & 9 & 6 \\ 2 & 3 & 5 & 4 \\ 1 & 2 & 3 & 5 \end{pmatrix}$$

Transform matrix to upper triangular form, using elementary row operations and properties of a matrix determinant.

$$\det \mathbf{A} = \begin{vmatrix} 1 & 2 & 4 & 5 \\ 8 & 5 & 9 & 6 \\ 2 & 3 & 5 & 4 \\ 1 & 2 & 3 & 5 \end{vmatrix}$$

$R_2 - 8\,R_1 \rightarrow R_2$ (multiply 1 row by 8 and subtract it from 2 rows); $R_3 - 2\,R_1 \rightarrow R_3$ (multiply 1 row by 2 and subtract it from 3 rows); $R_4 - 1\,R_1 \rightarrow R_4$ (multiply 1 row by 1 and subtract it from 4 row)

$$= \begin{vmatrix} 1 & 2 & 4 & 5 \\ 0 & -11 & -23 & -34 \\ 0 & -1 & -3 & -6 \\ 0 & 0 & -1 & 0 \end{vmatrix} =$$

$R_3 - \frac{1}{11} R_2 \rightarrow R_3$ (multiply 2 rows by $\frac{1}{11}$ and subtract it from 3 rows)

$$= \begin{vmatrix} 1 & 2 & 4 & 5 \\ 0 & -11 & -23 & -34 \\ 0 & 0 & -\frac{10}{11} & -\frac{32}{11} \\ 0 & 0 & -1 & 0 \end{vmatrix} =$$

$R_4 - 1.11\,R_3 \rightarrow R_4$ (multiply 3 rows by 1.1 and subtract it from 4 rows)

$$= \begin{vmatrix} 1 & 2 & 4 & 5 \\ 0 & -11 & -23 & -34 \\ 0 & 0 & -\frac{10}{11} & -\frac{32}{11} \\ 0 & 0 & 0 & 3.2 \end{vmatrix} = 1(-11)\left(-\frac{10}{11}\right)3.2 = 32$$

8.1.2 Rank of a Matrix

Example:

$$\mathbf{A} = \begin{pmatrix} 5 & 7 & 6 \\ 3 & 4 & 2 \\ 6 & 5 & 6 \end{pmatrix}$$

Solution

$$A=\begin{pmatrix} 2 & 3 & 2 \\ 4 & 5 & 2 \\ 5 & 6 & 2 \end{pmatrix}$$

To calculate matrix rank transform matrix to upper triangular form, using elementary row operations.
$R_1/5 \rightarrow R_1$ (divide 1 row by 5)

$$\begin{pmatrix} 1 & 1.4 & 1.2 \\ 3 & 4 & 2 \\ 6 & 5 & 6 \end{pmatrix}$$

R2 − 3 R1 → R2 (multiply 1 row by 3 and subtract it from 2 rows);
R3 − 6 R1 → R3 (multiply 1 row by 6 and subtract it from 3 rows)

$$\begin{pmatrix} 1 & 1.4 & 1.2 \\ 0 & -0.2 & -1.6 \\ 0 & -3.4 & -1.2 \end{pmatrix}$$

$R_2/-0.2 \rightarrow R_2$ (divide the 2 row by −0.2)

$$\begin{pmatrix} 1 & 1.4 & 1.2 \\ 0 & 1 & 8 \\ 0 & -3.4 & -1.2 \end{pmatrix}$$

$R_3+3.4\ R_2 \rightarrow R_3$ (multiple 2 rows by 3.4 and add it to 3 rows)

$$\begin{pmatrix} 1 & 1.4 & 1.2 \\ 0 & 1 & 8 \\ 0 & 0 & 26 \end{pmatrix}$$

$R_3/26 \rightarrow R_3$ (divide the 3 rows by 26)

$$\begin{pmatrix} 1 & 1.4 & 1.2 \\ 0 & 1 & 8 \\ 0 & 0 & 1 \end{pmatrix}$$

Since there are three non-zero rows, then Rank(**A**) = 3

Inverse

Matrices have both additive and multiplicative inverses.

Additive Inverse: The additive inverse of a matrix A is another matrix B such that A+B=0.

In other words $b_{ij}=-a_{ij}$.

Normally it is denoted as –A.

Multiplicative Inverse: The multiplicative inverse is defined only for square matrices.

The multiplicative inverse of a square matrix A is a square matrix B such that A×B=B×A=I.

It is denoted as A^{-1}.

The multiplicative inverse exists only if det(A) has a multiplicative inverse in the corresponding set.

Residue Matrices: A residue matrix has a multiplicative inverse if the determinant of the matrix has a multiplicative inverse in Zn.

In other words, a residue matrix has a multiplicative inverse if gcd(det(A),n)=1.

Inverse of a Matrix using Gauss Jordan Elimination method using Identity matrix.

Example:

$$\mathbf{A} = \begin{pmatrix} 3 & 2 & 4 \\ 4 & 5 & 5 \\ 5 & 3 & 1 \end{pmatrix}$$

Adjoin the identity matrix onto the right of the original matrix so that you have A on the left side and the identity matrix on the right side. It will look like this:

$$\left(\begin{array}{ccc|ccc} 3 & 2 & 4 & 1 & 0 & 0 \\ 4 & 5 & 5 & 0 & 1 & 0 \\ 5 & 3 & 1 & 0 & 0 & 1 \end{array}\right)$$

Now find the inverse matrix. Using elementary row operations to transform the left side of the resulting matrix to the identity matrix.

$R_1/3 \to R_1$ (divide 1 row by 3)

$$\left(\begin{array}{ccc|ccc} 1 & \frac{2}{3} & \frac{4}{3} & \frac{1}{3} & 0 & 0 \\ 4 & 5 & 5 & 0 & 1 & 0 \\ 5 & 3 & 1 & 0 & 0 & 1 \end{array}\right)$$

$R_2 - 4R_1 \to R_2$ (multiply 1 row by 4 and subtract it from 2 rows); $R_3 - 5R_1 \to R_3$ (multiply 1 row by 5 and subtract it from 3 rows)

$R_2/\frac{7}{3} \to R_2$ (divide the 2 row by $\frac{7}{3}$)

$$\left(\begin{array}{ccc|ccc} 1 & \frac{2}{3} & \frac{4}{3} & \frac{1}{3} & 0 & 0 \\ 0 & 1 & -\frac{1}{7} & -\frac{4}{7} & \frac{3}{7} & 0 \\ 0 & -\frac{1}{3} & -\frac{17}{3} & -\frac{5}{3} & 0 & 1 \end{array}\right)$$

$R_1 - \frac{2}{3} R_2 \to R_1$ (multiply 2 rows by $\frac{2}{3}$ and subtract it from 1 row); $R_3 + \frac{1}{3} R_2 \to R_3$ (multiply 1 row by 5 and subtract it from 3 rows)

$$\left(\begin{array}{ccc|ccc} 1 & 0 & \frac{10}{7} & \frac{5}{7} & -\frac{2}{7} & 0 \\ 0 & 1 & -\frac{1}{7} & -\frac{4}{7} & \frac{3}{7} & 0 \\ 0 & 0 & -\frac{40}{7} & -\frac{13}{7} & \frac{1}{7} & 1 \end{array}\right)$$

$R_3/-\frac{40}{7} \to R_3$ (divide 3 rows by $-\frac{40}{7}$)

$$\left(\begin{array}{ccc|ccc} 1 & 0 & \frac{10}{7} & \frac{5}{7} & -\frac{2}{7} & 0 \\ 0 & 1 & -\frac{1}{7} & -\frac{4}{7} & \frac{3}{7} & 0 \\ 0 & 0 & 1 & 0.325 & -0.025 & -0.0175 \end{array}\right)$$

$R_1 - \frac{10}{7} R_3 \rightarrow R_1$ (multiply 3 rows by $\frac{10}{7}$ and subtract it from 1 row)

$R_2 + \frac{1}{7} R_3 \rightarrow R_2$ (multiply 3 rows by $\frac{1}{7}$ and add it to 2 rows)

$$\left(\begin{array}{ccc|ccc} 1 & 0 & 0 & 0.25 & -0.25 & 0.25 \\ 0 & 1 & 0 & -0.525 & 0.425 & -0.025 \\ 0 & 0 & 1 & 0.325 & -0.025 & -0.175 \end{array}\right)$$

Finally,

$$\mathbf{A}^{-1} = \begin{pmatrix} 0.25 & -0.25 & 0.25 \\ -0.525 & 0.425 & -0.025 \\ 0.325 & -0.025 & -0.175 \end{pmatrix}$$

8.1.3 Application of Matrices

Matrices are used in various fields of scientific areas. In the branch of physics, including classical mechanics, optics, electromagnetism, quantum mechanics, and quantum electrodynamics, matrices are used to study physical phenomena, such as the motion of rigid bodies.

In computer graphics, they are used to project a three-dimensional image onto a two-dimensional screen. In probability theory and statistics, stochastic matrices are used to describe sets of probabilities; for instance, they are used within the Page Rank algorithm that ranks the pages in a Google search.

Matrix calculus generalizes classical analytical notions such as derivatives and exponentials to higher dimensions. The graphics software uses the concept of a matrix to process linear transformations to render images.

8.1.4 Boolean Matrix Multiplication

Boolean matrix multiplication is one of the most fundamental problems in computer science. It has applications to triangle finding, transitive closure and parsing of context-free grammars. The best way to multiply two Boolean matrices is to treat them as integer matrices.

Algorithm

Let A,B: n×n be the Boolean matrices.

Let C be the product matrix A×B.

Now choose a suitable submatrix block size k;
For every K×K matrix P and every K×K matrix Q
    Product [P,Q] := P×Q'
    Sum[P,Q] := P+Q
[End of loop]

Example:

Given a boolean matrix mat[M][N] of size M X N, modify it such that if a matrix cell mat[i][j] is 1 (or true), then make all the cells of ith row and jth column as 1.

Programming Implementations

```
#include <bits/stdc++.h>

using namespace std;
#define R 3
#define C 4

void modifyMatrix(bool mat[R][C])
{
    bool row[R];
    bool col[C];

    int i, j;

    /* Initialize all values of row[] as 0 */
    for (i = 0; i < R; i++)
    {
    row[i] = 0;
    }

    /* Initialize all values of col[] as 0 */
    for (i = 0; i < C; i++)
    {
    col[i] = 0;
    }

    // Store the rows and columns to be marked as
    // 1 in row[] and col[] arrays respectively
    for (i = 0; i < R; i++)
```

```
    {
        for (j = 0; j < C; j++)
        {
            if (mat[i][j] == 1)
            {
                row[i] = 1;
                col[j] = 1;
            }
        }
    }

    // Modify the input matrix mat[] using the
    // above constructed row[] and col[] arrays
    for (i = 0; i < R; i++)
    {
        for (j = 0; j < C; j++)
        {
            if ( row[i] == 1 || col[j] == 1 )
            {
                mat[i][j] = 1;
            }
        }
    }
}

/* A utility function to print a 2D matrix */
void printMatrix(bool mat[R][C])
{
    int i, j;
    for (i = 0; i < R; i++)
    {
        for (j = 0; j < C; j++)
        {
            cout << mat[i][j];
        }
        cout << endl;
    }
}

int main()
{
    bool mat[R][C] = { {1, 0, 0, 1},
```

```
                                    {0, 0, 1, 0},
                                    {0, 0, 0, 0}};

        cout<<endl<<"Enter the matrix";

  for(int i=0;i<R;i++)
        for(int j=0;j<C;j++)
            cin>>mat[i][j];

      cout << "Input Matrix \n";
      printMatrix(mat);

      modifyMatrix(mat);

      printf("Matrix after modification \n");
      printMatrix(mat);

      return 0;
  }
```

```
output

The matrix
0 0 0
0 0 1
should be changed to following
0 0 1
1 1 1
Input Matrix
1 0 0 1
0 0 1 0
0 0 0 0
Matrix after modification
1 1 1 1
1 1 1 1
1 0 1 1
```

8.2 POLYNOMIALS

The polynomials can also be represented in the form of $GF(2^n)$.

A polynomial of degree (n-1) can be represented as

$$f(x) = a_{n-1}x^{n-1} + a_{n-2}x^{n-2} + \ldots\ldots + a_1x^1 + a_0x^0$$

where xi is the ith term and ai is the coefficient of ith term.

Example:

Represent the 8-bit word 11011011 using polynomial.

n – bit word 1 1 0 1 0 1 1 1

Polynomial $1x^7 + 1x^6 + 0x^5 + 1x^4 + 0x^3 + 1x^2 + 1x^1 + 1x^0$

$$= 1x^7 + 1x^6 + 1x^4 + 1x^2 + 1x^1 + 1x^0$$

$$= x^7 + x^6 + x^4 + x^2 + x + 1$$

Similarly one can also find the 8-bit word from the polynomial.

Polynomials representing n-bit words use two fields as GF(2) and GF(2^n).

Modulus:

In general, the addition operation with two polynomial creates a polynomial with a degree less than or equal to (n–1), but the multiplication operation will result in the polynomial with a degree greater than (n–1). Therefore, to maintain the terms in the range, we need to perform the modulus operation just implementing the concepts of modular arithmetic.

Prime Polynomial: A prime polynomial cannot be factorized into a polynomial with a degree less than n. Such polynomials are also referred to as irreducible polynomials.

Ex:

$$(x+1), x$$

$$(x^2+x+1)$$

$$(x^3+x^2+1), (x^3+x+1)$$

Addition Operation of polynomial in GF(2)

The addition operation of two polynomials is to add the coefficients of two polynomials.

Example:

Add (x5+x2 +x) and (x3+x2+1) in GF(2^8).

Solution

Since the addition operation is to be performed in GF(2^8), first we have to represent the two above polynomials in GF(2^8):

$$0x^7 + 0x^6 + 1x^5 + 0x^4 + 0x^3 + 1x^2 + 1x^1 + 0x^0$$

$$0x^7 + 0x^6 + 0x^5 + 0x^4 + 1x^3 + 1x^2 + 0x^1 + 1x^0$$

$$0x^7 + 0x^6 + 1x^5 + 0x^4 + 1x^3 + 0x^2 + 1x^1 + 1x^0$$

$$= x^5 + x^3 + x + 1$$

8.3 POLYNOMIAL AND FFT

A polynomial in the variable x over an algebraic field F is a representation of a function A(x) as a formal sum:

$$A(x) = \sum_{j=0}^{n-1} ajx^j$$

For addition of two polynomial, it is very simple, but for multiplication, it's very difficult and the terms will be

$$C(x) = \sum_{j=0}^{2n-2} c_j x^j$$

$$c_j = \sum_{k=0}^{j} a_k b_{j-k}$$

If we use the complex roots of unity, we can evaluate and interpolate polynomials in θ(nlogn) time.

A complex nth root of unity is a complex number w such that $w^n = 1$.

There are exactly n complex nth roots of unity: $e^{(2\Pi ik)/n}$ for $k = 0,1,2,\ldots,n-1$.

To interrupt this formula, we will use the definition of the exponential of a complex number:

$$e^{iu} = \cos(u) + I\sin(u)$$

$$\text{so } wn = e^{(2\Pi i)/n}.$$

By using a method Fast Fourier Transform which takes advantage of the special properties of the complex roots of unity, we can compute DFTn(a) in time θ(nlogn) as opposed to the $\theta(n^2)$ time of the straightforward method.

The FFT method employes a divide-and-conquer strategy, using the even and odd index coefficients of A(x) separately to define the two new polynomials A[0] and A[1] of degree bound n/2:

$$A[0]x = a0 + a2x + a4x2 + \ldots + an - 2xn/2 - 1$$

$$A[1]x = a1 + a3x + a5x2 + \ldots + an - 1xn/2 - 1$$

$$\text{So } A(x) = A[0]x2 + x\,A[1]x2$$

so that the problem of evaluating A(x) at wn0,wn1,....wnn-1 reduces to

1. Evaluating the degree bound n/2 polynomials A[0]x and A[1]x at the points
   (wn0)2,(wn1)2,...(wnn-1)2.
2. Now combine these to get the solution.

Recursive-FFT(A)

```
 1. n ←length[A]
 2. if n = 1
 3. then return A
 4. wn ←e(2Πi)/n
 5. w ←1
 6. A[0] ← (A0,A2,.........,An-2)
 7. A[1] ← (A1,A3,.........,An-1)
 8. Y[0] ←RECURSIVE-FFT(A[0])
 9. Y[1] ←RECURSIVE-FFT(A[1])
10. for k ←0 to (n/2) - 1
11. do Yk ←Yk0 + wYk1
12. Yk + (n/2) ← Yk[0] - wYk1
13. w ← wwn
14. return Y (y is used to be column vector)
```

8.4 QUESTIONS

8.4.1 Short Questions

1. what is polynomial?
2. Explain matrix.
3. Describe inverse.
4. What is multiplicative inverse?
5. Define residue matrices.
6. Describe the application of matrices.
7. Define Boolean matrix multiplication.
8. Define modulus.
9. What is prime polynomial?
10. Represent the 8-bit word 11011011 using polynomial.

8.4.2 Long Questions

1. Inverse of a matrix using Gauss Jordan Elimination method using identify matrix $A = \begin{pmatrix} 3 & 2 & 4 \\ 4 & 5 & 5 \\ 5 & 3 & 1 \end{pmatrix}$
2. Write the Boolean matrix multiplication algorithm.
3. Write the recursive-FFT algorithm with an example.
4. Explain polynomial and FFT with algorithm.
5. Compute the determinant of

$$A = \begin{pmatrix} 1 & 2 & 4 & 5 \\ 8 & 5 & 9 & 6 \\ 2 & 3 & 5 & 4 \\ 1 & 2 & 3 & 4 \end{pmatrix}$$

6. Explain operation with matrix.

7. Short notes on
   - Determinant
   - Inverse
8. Explain Gauss Jordan elimination.
9. Short note on the application of matrixes.
10. Short note on Boolean matrix multiplication and chain multiplication.

CHAPTER 9

Number Theoretic Algorithms

9.1 COMPUTATIONAL NUMBER THEORY

Cryptography is completely related to the area of mathematics such as number theory, linear algebra and algebraic structures.

- **Integer Arithmetic**

  The concept of Integer Arithmetic is implemented for modular arithmetic.

  In mathematics, we already know the set of Integers Z covers all the integers starting from $-\infty$ to ∞.

  $$Z = (\ldots -5, -4, -3, -2, -1, 0, 1, 2, 3, 4, 5\ldots)$$

- **Binary Operation**

  The binary operations are those operations that take two operands and result in a single value. In cryptography, we will deal with three common binary operations such as addition, subtraction, and multiplication.

  However, for the division operation, it will not fit because it will result in two values such as quotient and remainder.

- **Integer Division**

  In mathematics, we can represent the division operation in the form of

  $$A = Q * N + R$$

DOI: 10.1201/9781003093886-9

where

A: Dividend
Q: Quotient
N: Divisor
R: Remainder

For example: Divide 33 by 8,

Here A: 33 N: 8 Q: 4 R: 1

However, in computer, the quotient and remainder can be calculated by using two different operators.

In integer division, we will impose two restrictions:

1. Q must be greater than 0.
2. R must be greater than or equal to 0.

How these restrictions can be implemented with the integer divisions?

How R will be +ve

To maintain the above-said restrictions, one has to decrement the value of Q by 1 and add the value of N to R to make it positive.

Example:
$-255 = -23 * 11 + -2$
Here to make R as +ve
First, decrease the value of Q by 1 so –23 will turn to –24.
Add the value of N to R so $-2 + 11 = 9$ (+ve).

Divisibility

If A is not zero but $R = 0$, then we can say that it is the condition of divisibility.

The relation can be represented as

$$A = Q * N \quad (\text{N is a Divisor of A})$$

Symbolically, it can be represented as
N|A (If Divisible).

$N \nmid A$ (If not Divisible)

Example: 5|55, 6|18, -6|24, $5 \nmid 13$, $3 \nmid 58$

Properties of Divisibility

Property – 1: If A|1, then A=±1.

Property – 2: If A|B and B|A, then A=±B.

Property – 3: If A|B and B|C, then A|C.

Property – 4: If A|B and A|C, then A | (m * B+n*C), where m and n are arbitrary integers.

A +ve integer can have more than one divisor.

Two facts about the divisor of +ve integers.

Fact – 1: The integer 1 has only one divisor.

Fact – 2: Any +ve integer has at least two divisors, 1 and itself (but it can have more).

9.2 GREATEST COMMON DIVISOR

Greatest Common Divisor (GCD) of two integers always needed for cryptography. Two +ve integers can have more common divisors but have only one GCD.

For example, the common divisors of 12 and 140 are 1, 2 and 4. But the GCD is 4.

Euclidean Algorithm for GCD

The Euclidean Algorithm based on the following two factors:

1. GCD(a,0)=a.
2. GCD(a,b)=GCD(b,r) where r is the remainder of a and b.

Example:

Find the GCD of 25 and 60

Q	a	b	r
0	25	60	25
2	60	25	10
2	25	10	5
2	10	5	0
	5	0	
	Greatest common divisor=5		

Extended Euclidean Algorithm

Given two integers a and b, we often need to find other two integers, s and t such that

$$s * a + t * b = gcd(a, b)$$

Algorithm

Step 1 r1 = a & r2 = b
Step 2 s1 = 1 & s2 = 0
Step 3 t1 = 0 & t2 = 1
Step 4 while(r2 > 0)

```
{
q = r1/r2;
r = r1 – q & r2
r1 = r2 & r2 = r
s = s1 – q * s2
s1 = s2 & s2 = s
t = t1 – q * t2
t1 = t2 & t2 = t
}
```

Step 5 GCD = r1
Step 6 s = s1
Step 7 t = t1

Example:

Find the GCD of 161 and 28 and also find the value of s and t.

q	r1	r2	r	s1	s2	s	t1	t2	t
5	161	28	21	1	0	1	0	1	–5
1	28	21	7	0	1	–1	1	–5	6
3	21	7	0	1	–1	4	–5	6	–23
	7	0		**–1**	4		**6**	–23	

Greatest Common Divisor = 7 s = –1 & t = 6.

According to the Extended Euclidean Algorithm, the relation will be

$$\mathbf{s * a + t * b = gcd(a, b) => (-1) * 161 + 6 * 28 = -161 + 168 = 7(GCD)}$$

9.3 LINEAR DIOPHANTINE EQUATIONS

The Euclidean Extended algorithm is used to solve the Linear Diophantine Equations (LDE). In mathematics, the LDE are represented as

$$ax + by = c$$

Here, we have to find the values of x and y, which will satisfy the equation.

These types of equations have either no solution or an infinite number of solutions.

Let us consider $d = GCD(a, b)$

If $d \nmid c$ then the equation has no solution.

If $d|c$, then the equation has an infinite number of solutions.

One of them is called ***particular*** and the rest are ***general***.

Particular Solution

If $d|c$, a particular solution to the above equation can be found using the following steps:

1. Reduce the equation to $\mathbf{a_1x + b_1y = c1}$ by dividing both sides of the equation by d.
2. Solve for s and t in the relation $\mathbf{a_1s + b_1t = 1}$ using the extended Euclidean algorithm.
3. Particular solution : $X_0 = (c/d)*s$ and $Y_0 = (c/d)*t$.

General Solutions

After finding the particular solution, the general solution can be found.

General Solution: $X = X0 + k(b/d)$ and $Y = Y0 - k(b/d)$ where k is integer.

Example:

Find the particular and general solution to the equation
$21x + 14y = 35$

Solution:

Here a =21, b = 14 and c = 35
$d = GCD(21, 14) = 7$

Since 7|35, so this equation has an infinite number of solutions.
So reduce the equation by dividing by d as

$$3x+2y=5$$

Now find the value of s and t from the equation $3s+2t=1$ by using the extended Euclidean algorithm.

The value of $s=1$ and $t=-1$

So the solutions are

Particular: $X0=(c/d) * s=(35/7) * 1=5$ and $Y0=(c/d)*t=(35/7) * (-1)=-5$

General: $X=X0+k(b/d)=5+k * (14/7)=5+k * (2)$

and $Y=Y0-k * (a/d)=(-5)-k * (21/7)=-5-k * 3$

where k is integer

Therefore the solutions are **(5, −5)**, (7, −8), (9, −11), (11, −14).

9.4 MODULAR ARITHMETIC

In modular arithmetic, we are concerned about the remainder of the integer division. This means that we can change the division relation ($a=q * n+r$) as with two inputs a and n and output r.

To find out this another operator is used known as modulo operator and is denoted as ***mod***. The input n is known as ***modulus***, and the output r is known as ***residue***.

Example:

a. 15 mod 4 = 3
b. 24 mod 8 = 0
c. −18 mod 14 = 10
[−18 mod 14 = −4 but since remainder should not be −ve so add −4 + 14 = 10]
d. −7 mod 10 = 3

Set of Residues: Zn

We already know that the result of a mod n will be less than n i.e./from 0, 1, 2, …, n − 1. In other words, we can say that the modulo operation creates a set, and the set in modular arithmetic is known as

Set of least residues modulo n or Zn

For example:

$Z_2 = \{0, 1\}$

$Z_5 = \{0, 1, 2, 3, 4\}$

$Z_9 = \{0, 1, 2, 3, 4, 5, 6, 7, 8\}$

Congruence

In cryptography, the term congruence is most commonly used which is commonly known as equality.

In mathematics, the mapping from Z to Zn is not one to one.

Infinite members of Z can map to one member of Zn.

For example:

$2 \bmod 10 = 2$

$12 \bmod 10 = 2$

$22 \bmod 10 = 2$ and so on.

The above relation in cryptography is known as congruence that means the numbers like 2, 12 and 22 are called congruent mod 10.

The congruence operator is represented as $(\equiv)$.

9.5 LINEAR CONGRUENCE

9.5.1 Single-Variable Linear Equations

Here we will observe how to solve the linear equations having a single variable.

Let consider the equations which are of the form $ax \equiv b \pmod{n}$.

The above equation might have no solution or a limited number of solutions.

To know how many solutions these equations have first we have to find out gcd (a, n).

Let gcd (a, n) = d.

$$\text{If } d \nmid b, \text{ then there is no solution}$$

If $d|b$, then there is d number of solutions.

Process to find the solutions

1. Reduce the equation by dividing both sides of the equation(including the modulus) by d.
2. Multiply both sides of the reduced equation by the multiplicative inverse of a to find the particular solution X_0.
3. The general solutions are $X=X_0+k(n/d)$ for $K=0, 1, 2, \ldots, (d-1)$.

Example:

Solve the equation $10x \equiv 2 \pmod{15}$.

Ans:

Gcd (10 and 15) = 5 and since 5 does not divide 2, so the equation has no solutions.

Example:

Solve the equation $14x \equiv 12 \pmod{18}$.

Ans:

Gcd (14, 18) = 2 since 2 divides 12 so we have exactly two solutions.

1. Reduce the equation as $7x \equiv 6 \pmod{9}$.
2. Multiply both sides of the equation by inverse of a
   i.e./$x \equiv 6(7^{-1}) \pmod{9}$
   $X_0=(6(7^{-1}) \pmod{9})=(6\times4) \pmod{9}=\mathbf{6}$.
3. $X_1=X_0+1\times(18/2)=\mathbf{15}$.

9.5.2 Set of Linear Equations

We can also solve a set of linear equations with the same modulus if the matrix formed from the coefficients of the variables is invertible.

For this, we will use three matrices where the first matrix will be from coefficients of variables. The second is a column matrix made from the variables. The third is a column matrix made from the values at the right-hand side of the congruence operator.

Let consider equations as

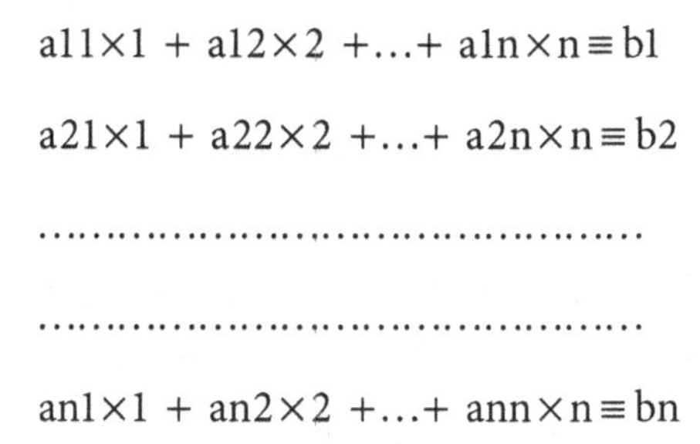

$$a11 \times 1 + a12 \times 2 + \ldots + a1n \times n \equiv b1$$

$$a21 \times 1 + a22 \times 2 + \ldots + a2n \times n \equiv b2$$

$$\ldots\ldots\ldots\ldots\ldots\ldots\ldots\ldots\ldots\ldots$$

$$\ldots\ldots\ldots\ldots\ldots\ldots\ldots\ldots\ldots\ldots$$

$$an1 \times 1 + an2 \times 2 + \ldots + ann \times n \equiv bn$$

Now the interpretation of the above equation will be

$$\begin{pmatrix} a_{11} & a_{12} & \cdots & a_{1n} \\ a_{21} & a_{22} & \cdots & a_{1n} \\ \cdots & \cdots & \cdots & \cdots \\ \cdots & \cdots & \cdots & \cdots \\ a_{n1} & a_{n2} & \cdots & a_{nn} \end{pmatrix} \begin{pmatrix} x_1 \\ x_2 \\ \cdots \\ \cdots \\ x_n \end{pmatrix} \equiv \begin{pmatrix} b_1 \\ b_2 \\ \cdots \\ \cdots \\ b_n \end{pmatrix}$$

The solution will be

$$\begin{pmatrix} x_1 \\ x_2 \\ \cdots \\ \cdots \\ x_n \end{pmatrix} \equiv \begin{pmatrix} a_{11} & a_{12} & \cdots & a_{1n} \\ a_{21} & a_{22} & \cdots & a_{1n} \\ \cdots & \cdots & \cdots & \cdots \\ \cdots & \cdots & \cdots & \cdots \\ a_{n1} & a_{n2} & \cdots & a_{nn} \end{pmatrix}^{-1} \begin{pmatrix} b_1 \\ b_2 \\ \cdots \\ \cdots \\ b_n \end{pmatrix}$$

9.6 GROUPS

A group is a set of elements with a binary operation "●" that satisfies four properties.

1. **Closure**: If a and b are elements of G, then $c = a \cdot b$ is also an element of G. This means that the result of applying the operation on any two elements in the set is another element in the set.

2. **Associativity**: If a, b and c are elements of G, then

$$(a \cdot b) \cdot c = a \cdot (b \cdot c).$$

That means it does not matter in which order we are applying the operations.

3. **Existence of Identity**: For all a in G, there exists an element e, called an identity element such that $e \cdot a = a \cdot e = a$.

4. **Existence of Inverse**: For each a in G, there exists an element a′ called the inverse of a such that $a \cdot a' = a' \cdot a = e$.

A ***commutative group/Abelian group*** is a group that will satisfy all the above properties plus another extra property Commutativity

Commutativity: For all a and b in G, we have $a \cdot b = b \cdot a$.

Finite Group:

A group is called a finite group if the set has a finite number of elements otherwise infinite group.

Order of a Group:

The order of a group is represented as |G| is the number of elements in the group. If the group is finite, its order is finite otherwise infinite.

Subgroups:

A subset H of a group G is a subgroup of G if H itself is a group with respect to the operation on G.

In other words, if G=<S, ·> is a group, H=<T, ·> is a group under the same operation and T is a non-empty subset of S, then H is a subgroup of G. The above definition implies that

1. If a and b are members of both groups, then $c = a \cdot b$ is also a member of both groups.

2. The group shares the same identity element.

3. If a is a member of both groups, the inverse of a is also a member of both groups.

4. The group made of the identity element of G, H=<{e}, ·>, is a subgroup of G.
5. Each group is a subgroup of itself.

Cyclic Subgroup:

If a subgroup of a group can be generated using the power of an element, the subgroup is called cyclic subgroup. The term power means repeated applying the group operation to the element.

That means $a^n = a \cdot a \cdot a \cdot a \ldots a$ (n times).

The duplicate elements are not included in this group and are represented as <a>.

Cyclic Groups:

A cyclic group is a group that is its own cyclic subgroup. In this case, the elements that generate the cyclic subgroup can also generate the group itself.

This element is referred to as a ***generator***.

If g is a generator then the elements in a finite cyclic group can be written as $\{e, g, g^2, g^3, \ldots, g^{n-1}\}$, where $g^n = e$.

A cyclic group can have many generators.

9.7 RING

A ring is denoted as R=<{...}, ·, □>, is an algebraic structure with two operations.

The first operation must have to satisfy all the five properties required for the Abelian group.

The second operation will satisfy only the first two.

The second operation is distributed over first.

Distributive means that for all a, b and c elements of R, we have

$$a \square (b \cdot c) = (a \square b) \cdot (a \square c) \text{and } (a \cdot b) \square c = (a \square c) \cdot (b \square c).$$

A commutative ring is a ring in which the commutative property is also for the second operation.

9.8 FIELD

A field, denoted as R=<{...}, ·, □> is a commutative ring in which the second operation satisfies all the five properties defined for the first operation except that the identity of the first operation (sometimes called the zero element) has no inverse.

Finite Fields:

A field with a finite number of elements called as finite fields.

A Galois field, $GF(P^n)$ is a finite field with P^n elements.

GF(P) Fields

A Galois field, $GF(P^n)$ is a finite field with P^n elements.

When the value of n is 1 then the Galois field can be called as GF(P) field.

The field can be the set $Z_p = \{0, 1, 2, 3, \ldots, p-1\}$ with two arithmetic operations such as addition and multiplication.

Example:

Perform the operations in GF(2).

The GF(2) field having 2 elements as {0, 1}

The operations to perform are addition and multiplication.

GF(2)

{0, 1}	+, x

Addition

+	0	1
0	0	1
1	1	0

Multiplication

x	0	1
0	0	0
1	0	1

Additive inverse

a	0	1
−a	0	1

Multiplicative inverse

a	0	1
a^{-1}	---	1

In GF(2), the addition operation results like XOR operation and multiplication like AND operations.

Example:

Perform the operations in GF(5).

The GF(5) field having five elements as {0, 1, 2, 3, 4}

The operations to perform are addition and multiplication.

GF(5)

{0, 1, 2, 3, 4}	+, ×

Addition

a	0	1	2	3	4
−a	0	4	3	2	1

Multiplication

+	0	1	2	3	4
0	0	1	2	3	4
1	1	2	3	4	0
2	2	3	4	0	1
3	3	4	0	1	2
4	4	0	1	2	3

Additive inverse

x	0	1	2	3	4
0	0	0	0	0	0
1	0	1	2	3	4
2	0	2	4	1	3
3	0	3	1	4	2
4	0	4	3	2	1

Multiplicative inverse

a	0	1	2	3	4
a^{-1}	...	1	3	2	4

Like GF(2), GF(5), we can also define $GF(P^n)$ fields.

$GF(2^n)$ Fields

In general, in the process of cryptography, we often use the four operations:

- Addition
- Subtraction
- Multiplication
- Division

In computer, the positive integers are stored as n-bit words in which n is usually 8, 16, 32, 64, etc.

That means the range of integers is 0 to 2^n-1 because the modulus is 2^n.

If we want to use the $GF(2^n)$ field, then we have two choices:

1. We can use GF(P) with the set Z_p where p is the largest prime number less than 2^n. It is inefficient because we cannot use the integers from p to 2^{n-1}.

   For example, if n=4, then the largest prime number less than 2^4 is 13. Therefore, we cannot use the integers 13, 14 and 15.

2. We can work in $GF(2^n)$ and uses a set of 2^n elements. The elements in this set are n-bit words. For example, if n=3, the set is
   {000, 001, 010, 011, 100, 101, 110, 111}

Example:

Perform the operations in $GF(2^2)$ field with 2-bit words.

Solution:

The elements are {00, 01, 10, 11}

Addition

+	00	01	10	11
00	**00**	**01**	**10**	**11**
01	**01**	**00**	**11**	**10**
10	**10**	**11**	**00**	**01**
11	**11**	**10**	**01**	**00**

Multiplication

×	00	01	10	11
00	**00**	**00**	**00**	**00**
01	**00**	**01**	**10**	**11**
10	**00**	**10**	**11**	**01**
11	**00**	**11**	**01**	**10**

9.9 QUESTIONS

9.9.1 Short Questions

1. What is integer arithmetic?
2. Explain binary operation.
3. Describe integer division.
4. What is divisibility?
5. Describe properties of divisibility.
6. Explain GCD.
7. Explain LDE.
8. What is a particular solution?
9. What is congruence?
10. Describe Ring and Field.

9.9.2 Long Questions

1. Perform the operations in GF(5).
2. Perform the operations in GF(2^2) field with 2-bit words.
3. Briefly explain GF(2^n) fields.

4. Differentiate between GF(2) and GF(5).
5. Describe cyclic subgroups and cyclic groups.
6. Short notes on
    - Ring
    - Field
7. Define commutative group/Abelian group.
8. Briefly discuss groups.
9. Solve the equation $14\times \equiv 12 (\text{mod } 18)$.
10. Discuss linear congruence.

CHAPTER 10

Programming Implementations of the Algorithms

10.1 PROGRAM FOR THE LONGEST COMMON SUBSEQUENCES

```
#include<stdio.h>
#include<string.h>
#define maxn 100100
int max(int a,int b)
{
        return a>b?a:b;
}
int LongestCommonSubsequence(char S[],char T[])
{
        int Slength = strlen(S);
        int Tlength = strlen(T);
        /* Starting the index from 1 for our
        convenience (avoids handling special cases for
        negative indices) */
        int iter,jter;
        for(iter=Slength;iter>=1;iter--)
        {
                S[iter] = S[iter-1];
        }
```

DOI: 10.1201/9781003093886-10

```
for(iter=Tlength;iter>=1;iter--)
{
        T[iter] = T[iter-1];
}
int common[Slength+1][Tlength+1];
/* common[i][j] represents length of the
longest common sequence in S[1..i], T[1..j] */
/* Recurrence: common[i][j] = common[i-1]
[j-1] + 1 if S[i]==T[j]
                              = max(common[i-1]
                              [j],common[i]
                              [j-1]) otherwise
*/
/*common[0][i]=0, for all i because there are
no characters from string S*/
for(iter=0;iter<=Tlength;iter++)
{
        common[0][iter]=0;
}
/*common[i][0]=0, for all i because there are
no characters from string T*/
for(iter=0;iter<=Slength;iter++)
{
        common[iter][0]=0;
}
for(iter=1;iter<=Slength;iter++)
{
        for(jter=1;jter<=Tlength;jter++)
        {
                if(S[iter] == T[jter] )
                {
                        common[iter][jter] =
                        common[iter-1][jter-1]
                        + 1;
                }
                else
                {
                        common[iter][jter] =
                        max(common[iter][jter-
                        1],common[iter-1]
                        [jter]);
```

```
                    }
                }
        }
        return common[Slength][Tlength];

}
int main()
{
        char S[maxn],T[maxn];/* S,T are two strings
        for which we have to find the longest common
        sub sequence. */
        scanf("%s%s",S,T);
        printf("%d\n",LongestCommonSubsequence(S,T));
}
```

10.2 MATRIX CHAIN MULTIPLICATION

```
#include<stdio.h>
#include<conio.h>
#define MAX 10
void print(int s[MAX][MAX],int,int);
void main()
 {
   int i,j,q,k,l,n;
   int a[MAX][MAX]={0};
   int s[MAX][MAX]={0};
   int p[MAX];
   clrscr();
   printf("\nEnter No. of matrix");
   scanf("%d",&n);
   printf("\nEnter matrix %d diamentions\n",n+1);
   for(i=0;i<=n;i++)
    {
     scanf("%d",&p[i]);
    }

  for(l=2;l<=n;l++)
  {
    for(i=1;i<=n-l+1;i++)
    {
       j=i+l-1;
```

```
            a[i][j]='/0';
            for(k=i;k<=j-1;k++)
            {
        q=(a[i][k])+(a[k+1][j])+(p[i-1]*p[k]*p[j]);
        if(q<a[i][j])
          {
           a[i][j]=q;
           s[i][j]=k;
          }
        }

          }
      }
      printf("\nSequence is :\n");
      print(s,i-1,j);
      printf("\n\n");
      for(i=1;i<=n;i++)
       {
        for(j=1;j<=n;j++)
        {
        printf(" %5d",a[i][j]);
        }
        printf("\n");
       }
       printf("\nTotal required multiplication is :
       %d\n",a[1][n]);
       getch();
  }
  void print(int s[][MAX], int i, int j)
  {
  if (i == j)
  printf(" A%d ",i);
  else
  {
  printf(" ( ");
  print(s, i, s[i][j]);
  print(s, s[i][j] + 1, j);
  printf(" ) ");
  }
  }
```

10.3 PROGRAM FOR KNAPSACK PROBLEM

```
# include<stdio.h>
# include<conio.h>

void knapsack(int n, float weight[], float profit[],
float capacity)
{
 float x[20], tp= 0;
 int i, j, u;
 u=capacity;

 for (i=0;i<n;i++)
     x[i]=0.0;

 for (i=0;i<n;i++)
 {
 if(weight[i]>u)
      break;
 else
     {
     x[i]=1.0;
     tp= tp+profit[i];
     u=u-weight[i];
     }
 }

 if(i<n)
       x[i]=u/weight[i];

 tp= tp + (x[i]*profit[i]);

 printf("n The result vector is:- ");
 for(i=0;i<n;i++)
        printf("%ft",x[i]);

 printf("m Maximum profit is:- %f", tp);

}

void main()
{
 float weight[20], profit[20], capacity;
```

```
int n, i ,j;
float ratio[20], temp;
clrscr();

printf ("n Enter the no. of objects:- ");
scanf ("%d", &num);

printf ("n Enter the wts and profits of each
object:- ");
for (i=0; i<n; i++)
{
scanf("%f %f", &weight[i], &profit[i]);
}

printf ("n enter the capacity of knapsack:- ");
scanf ("%f", &capacity);

for (i=0; i<n; i++)
{
ratio[i]=profit[i]/weight[i];
}

for(i=0; i<n; i++)
{
   for(j=i+1;j< n; j++)
   {
     if(ratio[i]<ratio[j])
     {
     temp= ratio[j];
     ratio[j]= ratio[i];
     ratio[i]= temp;

    temp= weight[j];
    weight[j]= weight[i];
    weight[i]= temp;

    temp= profit[j];
    profit[j]= profit[i];
    profit[i]= temp;
    }
  }
}
```

```
 knapsack(n, weight, profit, capacity);
 getch();
}
```

Output:

```
Enter the no. of objects:- 7

Enter the wts and profits of each object:-
2 10
3 5
5 15
7 7
1 6
4 18
1 3

Enter the capacity of knapsack:- 15

The result vector is:- 1.000000        1.000000
1.000000        1.000000
    1.000000        0.666667        0.000000

Maximum profit is:-  55.333332
```

10.4 BELLMAN–FORD PROGRAM

```
// A C / C++ program for Bellman-Ford's single source
shortest path algorithm.

#include <stdio.h>
#include <stdlib.h>
#include <string.h>
#include <limits.h>

// a structure to represent a weighted edge in graph
struct Edge
{
    int src, dest, weight;
};

// a structure to represent a connected, directed and
weighted graph
```

```
struct Graph
{
    // V-> Number of vertices, E-> Number of edges
    int V, E;

    // graph is represented as an array of edges.
    struct Edge* edge;
};

// Creates a graph with V vertices and E edges
struct Graph* createGraph(int V, int E)
{
    struct Graph* graph = (struct Graph*) malloc(
    sizeof(struct Graph) );
    graph->V = V;
    graph->E = E;

    graph->edge = (struct Edge*) malloc( graph->E *
    sizeof( struct Edge ) );

    return graph;
}

// A utility function used to print the solution
void printArr(int dist[], int n)
{
    printf("Vertex   Distance from Source\n");
    for (int i = 0; i < n; ++i)
        printf("%d \t\t %d\n", i, dist[i]);
}

// The main function that finds shortest distances
from src to all other
// vertices using Bellman-Ford algorithm.  The
function also detects negative
// weight cycle
void BellmanFord(struct Graph* graph, int src)
{
    int V = graph->V;
    int E = graph->E;
    int dist[20];
```

```
    // Step 1: Initialize distances from src to all
    other vertices as INFINITE
    for (int i = 0; i < V; i++)
        dist[i]   = INT_MAX;
    dist[src] = 0;

    // Step 2: Relax all edges |V| - 1 times. A simple
    shortest path from src
    // to any other vertex can have at-most |V| - 1 edges
    for ( i = 1; i <= V-1; i++)
    {
        for (int j = 0; j < E; j++)
        {
            int u = graph->edge[j].src;
            int v = graph->edge[j].dest;
            int weight = graph->edge[j].weight;
            if (dist[u] + weight < dist[v])
                dist[v] = dist[u] + weight;
        }
    }

    // Step 3: check for negative-weight cycles.  The
    above step guarantees
    // shortest distances if graph doesn't contain
    negative weight cycle.
    // If we get a shorter path, then there is a cycle.
    for ( i = 0; i < E; i++)
    {
        int u = graph->edge[i].src;
        int v = graph->edge[i].dest;
        int weight = graph->edge[i].weight;
        if (dist[u] + weight < dist[v])
            printf("Graph contains negative weight
            cycle");
    }

    printArr(dist, V);

    return;
}

// Driver program to test above functions
```

```
int main()
{
    /* Let us create the graph given in above
    example */
    int V = 5;  // Number of vertices in graph
    int E = 8;  // Number of edges in graph
    struct Graph* graph = createGraph(V, E);

    // add edge 0-1 (or A-B in above figure)
    graph->edge[0].src = 0;
    graph->edge[0].dest = 1;
    graph->edge[0].weight = -1;

    // add edge 0-2 (or A-C in above figure)
    graph->edge[1].src = 0;
    graph->edge[1].dest = 2;
    graph->edge[1].weight = 4;

    // add edge 1-2 (or B-C in above figure)
    graph->edge[2].src = 1;
    graph->edge[2].dest = 2;
    graph->edge[2].weight = 3;

    // add edge 1-3 (or B-D in above figure)
    graph->edge[3].src = 1;
    graph->edge[3].dest = 3;
    graph->edge[3].weight = 2;

    // add edge 1-4 (or A-E in above figure)
    graph->edge[4].src = 1;
    graph->edge[4].dest = 4;
    graph->edge[4].weight = 2;

    // add edge 3-2 (or D-C in above figure)
    graph->edge[5].src = 3;
    graph->edge[5].dest = 2;
    graph->edge[5].weight = 5;

    // add edge 3-1 (or D-B in above figure)
    graph->edge[6].src = 3;
    graph->edge[6].dest = 1;
    graph->edge[6].weight = 1;
```

```
    // add edge 4-3 (or E-D in above figure)
    graph->edge[7].src = 4;
    graph->edge[7].dest = 3;
    graph->edge[7].weight = -3;

    BellmanFord(graph, 0);

    return 0;
}
```

Output:

```
Vertex   Distance from Source
0                0
1                -1
2                2
3                -2
4                1
```

10.5 WRITE A PROGRAM FOR TRAVELLING SALESMAN PROBLEM USING BACKTRACKING

```
/* Travelling Salesman Problem using backtracking*/
#include"stdio.h"
int x[15],used[15];
int adj[15][15]={0};
int path[15][15],wght[15];
int c,min;
int path_ok(int k,int n)
{
if(used[x[k]])
return 0;
if(k<n-1)
return(adj[x[k-1]][x[k]]);
else
return(adj[x[k-1]][x[k]] && adj[x[k]][x[0]]);
}
void TSP(int k,int n)
{
int i,sum;
for(x[k]=1;x[k]<n;x[k]++)
{
```

```
if(path_ok(k,n))
{
used[x[k]]=1;
if(k==n-1)
{
sum=0;
printf("\n\n\tPOSSIBLE PATH %d : ",c+1);
for(i=0;i<n;i++)
{
printf("%d\t",x[i]);
path[c][i]=x[i];
sum+=adj[x[i]][x[i+1]];
}
printf(" : %d",sum);
wght[c]=sum;
if(c==0 || sum<min)
min=sum;
c++;
used[x[k]]=0;
getch();
}
else
TSP(k+1,n);
used[x[k]]=0;
}
}
}
void findmin(int n)
{
int i,j;
for(i=0;i<c;i++)
if(wght[i]==min)
{
printf("\n\n\tMINIMUM PATH : ");
for(j=0;j<n;j++)
printf("%d\t",path[i][j]);
}
}
void main()
{
int i,n,j;
int edg;
```

```
clrscr();
printf("\n\n\t\tTRAVELLING SALESMAN PROBLEM\n\n");
printf("\n\tEnter the no. of Cities : ");
scanf("%d",&n);
printf("\n\n Enter the Cost if path Exist Between
cities.:{c1,c2}.Else Enter 0\n\n");
printf("\n\tCITIES\t\tCOST\n\n");
for(i=0;i<n;i++)
for(j=i+1;j<n;j++)
{
printf("\n\t %d------ %d \t:",i,j);
scanf("%d",&edg);
if(edg)
adj[i][j]=adj[j][i]=edg;
}
used[0]=1;
TSP(1,n);
if(!c)
printf("\n\n\t\tNO PATH FOUND TO COVER ALL THE
CITIES\n\n");
else
{
printf("\n\n\t\tMINIMUM COST IS %d AND THE PATHS ARE
\n\n",min);
findmin(n);
}
getch();
}
```

10.6 TRAVELING SALESMAN PROBLEM USING BRANCH AND BOUND

```
/*Branch and Bound Algorithm for Travelling Sales
Person*/
#include<stdio.h>
#include<conio.h>
int a[10][10],visited[10],n,cost=0;

void get()
{
int i,j;
printf("Enter No. of Cities: ");
```

```
scanf("%d",&n);
printf("\nEnter Cost Matrix: \n");
for( i=0;i<n;i++)
{
printf("\n Enter Elements of Row # : %d\n",i+1);
for( j=0;j<n;j++)
scanf("%d",&a[i][j]);
visited[i]=0;
}
printf("\n\nThe cost list is:\n\n");
for( i=0;i<n;i++)
{
printf("\n\n");
for( j=0;j<n;j++)
printf("\t%d",a[i][j]);
}
}

void mincost(int city)
{
int i,ncity;
visited[city]=1;
printf("%d ->",city+1);
ncity=least(city);
if(ncity==999)
{
ncity=0;
printf("%d",ncity+1);
cost+=a[city][ncity];
return;
}
mincost(ncity);
}
int least(int c)
{
int i,nc=999;
int min=999,kmin;
for(i=0;i<n;i++)
{
if((a[c][i]!=0)&&(visited[i]==0))
if(a[c][i]<min)
{
```

```
min=a[i][0]+a[c][i];
kmin=a[c][i];
nc=i;
}
}
if(min!=999)
cost+=kmin;
return nc;
}

void put()
{
printf("\n\nMinimum cost:");
printf("%d",cost);
}
void main()
{
clrscr();
get();
printf("\n\nThe Path is:\n\n");
mincost(0);
put();
getch();
}
```

Input Sample:
No. of Nodes: 6

99	10	15	20	99	8
5	99	9	10	8	99
6	13	99	12	99	5
8	8	9	99	6	99
99	10	99	6	99	99
10	99	5	99	99	99

Cost Matrix:

10.7 PROGRAM FOR HEAP SORT

```
#include<stdio.h>
#include <time.h>
```

```
#include<stdlib.h>
void heapsort(int[],int);
void heapify(int[],int);
void adjust(int[],int);
//driver program
main()
{
int n,i,a[500];
clock_t start, end;
double time_used=0;

printf("\nEnter the number of elements to sort:");
scanf("%d",&n);
printf("\nEnter %d elements:",n);
for (i=0;i<n;i++)
{
printf("\n Enter %d element",i+1);
scanf("%d",&a[i]);
}
 /* Recording the starting clock tick.*/
   start = clock();
heapsort(a,n);
  // Recording the end clock tick.
    end = clock();

  // Calculating total time taken by the program.
time_used = (double)(end - start) / (double)
(CLOCKS_PER_SEC);
 printf("Time required for sort %d elements is %lf
 secs.",n,time_used);
printf("\nThe Sorted Elements Are:\n");
for (i=0;i<n;i++)
  printf(" %d  ",a[i]);
//loop to generate 100 numbers and sort them using
heapsort
for(i=0;i<100;i++)
 {
a[i]= rand() % 500 +1;
 }
 time_used=0;
/* Recording the starting clock tick.*/
```

```
    start = clock();
heapsort(a,100);
// Recording the end clock tick.
    end = clock();

  // Calculating total time taken by the program.
time_used = (double)(end - start) / (double)
(CLOCKS_PER_SEC);
 printf("\nTime required for sort 100 elements is %lf
 secs.",time_used);
printf("\nThe Sorted Elements Are:\n");
for (i=0;i<100;i++)
  printf("\t%d",a[i]);
printf("\n");
}


//heapsort() method

void heapsort(int a[],int n)
{
int i,t;
heapify(a,n); //call to heapify method
for (i=n-1;i>0;i--) {
t = a[0];  //perform swap operation
a[0] = a[i];
a[i] = t;
adjust(a,i);
}
}
//heapify() method
void heapify(int a[],int n) {
int k,i,j,item;
for (k=1;k<n;k++) {
item = a[k];
i = k;
j = (i-1)/2;
while((i>0)&&(item>a[j])) {
a[i] = a[j];
i = j;
j = (i-1)/2;
}
```

```
a[i] = item;
}
}
//adjust() methjod
void adjust(int a[],int n) {
int i,j,item;
j = 0;
item = a[j];
i = 2*j+1;
while(i<=n-1) {
if(i+1 <= n-1)
   if(a[i] <a[i+1])
    i++;
if(item<a[i]) {
a[j] = a[i];
j = i;
i = 2*j+1;
} else
   break;
}
a[j] = item;
}
```

10.8 PROGRAM FOR QUICK SORT

```
#include <stdio.h>

void quick_sort(int[],int,int);
int partition(int[],int,int);

int main()
{
      int a[10],i;
      //input array elements
      printf("\nEnter array elements:");

      for(i=0;i<10;i++)
            scanf("%d",&a[i]);
            //call quick sort
      quick_sort(a,0,10-1);
      printf("\nArray after sorting:");
      //display the sorted array
```

```
	for(i=0;i<10;i++)
		printf("%d ",a[i]);

	return 0;
}
 //quick sort
void quick_sort(int a[],int l,int u)
{
	int j,i; static int count=0;
	if(l<u)
	{
		j=partition(a,l,u);
		count++; //count number of partitions
		printf("\n Partition %d : ",count);
		//display the array elements after each
		partition
		for(i=0;i<10;i++)
		printf("  %d  ",a[i]);
		quick_sort(a,l,j-1);
		quick_sort(a,j+1,u);
	}
}
 //partition method
int partition(int a[],int l,int u)
{
	int v,i,j,temp;
	v=a[l];
	i=l;
	j=u+1;
	do
	{
		do
			i++;

		while(a[i]<v&&i<=u);

		do
			j--;
		while(v<a[j]);

		if(i<j)
```

```
            {
                  temp=a[i];
                  a[i]=a[j];
                  a[j]=temp;
            }
      }while(i<j);

      a[l]=a[j];
      a[j]=v;

      return(j);
}
```

10.9 PROGRAM FOR MERGE SORT

```
#include <stdio.h>
   #include<iostream>
   #include<iomanip>
   using namespace std;
   // function to sort the  array a[] using merge sort
   void merge(int a[], int aux[],int lo,int mid,int hi)
   {
      if (hi <= lo) //condition for the subarray
      having no element or single element
         {
       return;
     }
     merge(a, aux,lo,(lo+mid)/2,mid);     // sort the
     left sub-array recursively
     merge(a, aux,mid+1,(mid+1+hi)/2,hi);     // sort
     the right sub-array recursively

     int low = lo;        // low points to the beginning
     of the left sub-array
     int high = mid + 1;        // high points to the
     beginning of the right sub-array
     int k;      // k is the loop counter

     for (k = lo; k <= hi; k++) {
```

```
        if (low == mid + 1) {      // left pointer has
        reached the limit
           aux[k] = a[high];
           high++;
       } else if (high == hi + 1) {        // right
       pointer has reached the limit
           aux[k] = a[low];
           low++;
       } else if (a[low] < a[high]) {      // pointer
       left points to smaller element
           aux[k] = a[low];
           low++;
       } else {     // pointer right points to smaller
       element
           aux[k] = a[high];
           high++;
       }
   }

   for (k = lo; k <= hi; k++) {   // copy the elements
   from aux[] to a[]
       a[k] = aux[k];
   }
}

int main() {
  int a[100], aux[100], n, i;

  printf("Enter number of elements in both the
  arrays:\n");
  scanf("%d", &n);

  printf("Enter %d integers for A followed by B arrays
  \n", n);

  for (i = 0; i < n; i++)
    scanf("%d", &a[i]);

  merge(a,aux,0,(0+n-1)/2,n - 1);

  cout<<endl<<"Elements after merge sort : ";
```

```
  for (i = 0; i < n; i++)
     cout<<setw(5)<<a[i];

  return 0;
}
```

10.10 PROGRAM FOR DFS

```
#include <stdio.h>
#include <stdlib.h>
enum color{White, Gray, Black};
/*
  Node for linked list of adjacent elements.
  This will contain a pointer for next node.
  It will not contain the real element but the index of
  element of the array containing all the vertices V.
*/
typedef struct list_node {
  int index_of_item;
  struct list_node *next;
}list_node;

/*
  Node to store the real element.
  Contain data and pointer to the
  first element (head) of the adjacency list.
*/
typedef struct node {
  int data;
  enum color colr;
  list_node *head;
}node;

/*
  Graph will contain number of vertices and
  an array containing all the nodes (V).
*/
typedef struct graph{
  int number_of_vertices;
  node heads[]; // array of nodes to store the list of
  first nodes of each adjacency list
```

```
}graph;

node* new_node(int data) {
  node *z;
  z = malloc(sizeof(node));
  z->data = data;
  z->head = NULL;
  z->colr = White;

  return z;
}

list_node* new_list_node(int item_index) {
  list_node *z;
  z = malloc(sizeof(list_node));
  z->index_of_item = item_index;
  z->next = NULL;

  return z;
}

// make a new graph
graph* new_graph(int number_of_vertices) {
  //number_of_vertices*sizeof(node) is the size of the
  array heads
  graph *g = malloc(sizeof(graph) + (number_of_vertices*
  sizeof(node)));
  g->number_of_vertices = number_of_vertices;

  //making elements of all head null i.e.,
  //their data -1 and next null
  int i;
  for(i=0; i<number_of_vertices; i++) {
    node *z = new_node(-1); //*z is pointer of node. z
    stores address of node
    g->heads[i] = *z; //*z is the value at the address z
  }

  return g;
}

// function to add new node to graph
```

```
void add_node_to_graph(graph *g, int data) {
  // creating a new node;
  node *z = new_node(data);
  //this node will be added into the heads array of
  the graph g
  int i;
  for(i=0; i<g->number_of_vertices; i++) {
    // we will add node when the data in the node is -1
    if (g->heads[i].data < 0) {
      g->heads[i] = *z; //*z is the value at the
      address z
      break; //node is added
    }
  }
}

// function to check of the node is in the head array
of graph or not
int in_graph_head_list(graph *g, int data) {
  int i;
  for(i=0; i<g->number_of_vertices; i++) {
    if(g->heads[i].data == data)
      return 1;
  }
  return 0;
}

// function to add edge
void add_edge(graph *g, int source, int dest) {
  //if source or edge is not in the graph, add it
  if(!in_graph_head_list(g, source)) {
    add_node_to_graph(g, source);
  }
  if(!in_graph_head_list(g, dest)) {
    add_node_to_graph(g, dest);
  }

  int i,j;
  // iterating over heads array to find the source node
  for(i=0; i<g->number_of_vertices; i++) {
    if(g->heads[i].data == source) { //source node
    found
```

```
      int dest_index; //index of destination element
      in array heads
      // iterating over heads array to find node
      containing destination element
      for(j=0; j<g->number_of_vertices; j++) {
        if(g->heads[j].data == dest) { //destination
        found
          dest_index = j;
          break;
        }
      }

      list_node *n = new_list_node(dest_index); // new
      adjacency list node with destination index
      if (g->heads[i].head == NULL) { // no head,
      first element in adjacency list
        g->heads[i].head = n;
      }
      else { // there is head which is pointer by the
      node in the head array
        list_node *temp;
        temp = g->heads[i].head;

        // iterating over adjacency list to insert new
        list_node at last
        while(temp->next != NULL) {
          temp = temp->next;
        }
        temp->next = n;
      }
      break;
    }
  }
}

void print_graph(graph *g) {
  int i;
  for(i=0; i<g->number_of_vertices; i++) {
    list_node *temp;
    temp = g->heads[i].head;
    printf("%d\t",g->heads[i].data);
    while(temp != NULL) {
```

```
      printf("%d\t",g->heads[temp->index_of_item].
      data);
      temp = temp->next;
    }
    printf("\n");
  }
}

void dfs_visit(graph *g, node *i) {
  i->colr = Gray;

  list_node *temp;
  temp = i->head;
  while(temp != NULL) {
    if (g->heads[temp->index_of_item].colr == White) {
      dfs_visit(g, &g->heads[temp->index_of_item]);
    }
    temp = temp->next;
  }
  i->colr = Black;
  printf("%d\n",i->data);
}

void dfs(graph *g) {
  int i;
  for(i=0; i<g->number_of_vertices; i++) {
    g->heads[i].colr = White;
  }

  for(i=0; i<g->number_of_vertices; i++) {
    if (g->heads[i].colr == White) {
      dfs_visit(g, &g->heads[i]);
    }
  }
}

int main() {
  graph *g = new_graph(7);
  add_edge(g, 1, 2);
  add_edge(g, 1, 5);
```

```
  add_edge(g, 1, 3);
  add_edge(g, 2, 6);
  add_edge(g, 2, 4);
  add_edge(g, 5, 4);
  add_edge(g, 3, 4);
  add_edge(g, 3, 7);
  dfs(g);
  return 0;
}
```

10.11 PROGRAM FOR PRIMS ALGORITHM

```
#include<stdio.h>
#include<conio.h>
int a,b,u,v,n,i,j,ne=1;
int visited[10]= {0},min,mincost=0,cost[10][10];
void main() {

printf("\n Enter the number of nodes:");
scanf("%d",&n);
printf("\n Enter the adjacency matrix:\n");
for (i=1;i<=n;i++)
 for (j=1;j<=n;j++) {
      scanf("%d",&cost[i][j]);
      if(cost[i][j]==0)
         cost[i][j]=999;
}
visited[1]=1;
printf("\n");
while(ne<n) {
      for (i=1,min=999;i<=n;i++)
        for (j=1;j<=n;j++)
         if(cost[i][j]<min)
          if(visited[i]!=0) {
      min=cost[i][j];
      a=u=i;
      b=v=j;
      }
      if(visited[u]==0 || visited[v]==0) {
      printf("\n Edge %d:(%d %d) cost:%d",ne++,a,b,min);
      mincost+=min;
```

```
      visited[b]=1;
      }
      cost[a][b]=cost[b][a]=999;
}
printf("\n Minimun cost=%d",mincost);
getch();
}
```

10.12 PROGRAM FOR THE WARSHALL METHOD

```
      #include<stdio.h>
int i, j, k,n,x,y,dist[10][10];
void floydWarshell ()
{
 for (k = 0; k < n; k++)
  {
      printf("\n For K = %d",k);
      for (i = 0; i < n; i++)
      {
      for (j = 0; j < n; j++)
      {
      if (dist[i][k] + dist[k][j] < dist[i][j])
      {
             printf("\nFor i=%d, j= %d,dist[%d]
             [%d]+dist[%d][%d] < dist[%d][%d], %d + %d
             < %d (T), dist[%d][%d] = %d",i,j,i,k,k,j,
             i,j,dist[i][k] ,dist[k][j],dist[i]
             [j],i,j,dist[i][k] + dist[k][j]);
            dist[i][j] = dist[i][k] + dist[k][j];
         }
         else
         {
             printf("\n For i=%d, j= %d,dist[%d][%d] +
             dist[%d][%d] < dist[%d][%d] i.e/ %d + %d
             < %d (False)",i,j,i,k,k,j,i,j,dist[i][k]
             ,dist[k][j],dist[i][j]);
         }

         }

      }
      printf("\n\n PATH MATRIX - %d\n",k+1) ;
```

```
        for (x = 0; x < n; x++)
      {
            for (y = 0; y < n; y++)
                    printf ("%d\t", dist[x][y]);
            printf("\n");
      }
   getch();

  }

}

int main()
{
  int i,j;
  printf("enter no of vertices :");
  scanf("%d",&n);
  printf("\n");
  for(i=0;i<n;i++)
  for(j=0;j<n;j++)
   {
    printf("dist[%d][%d]:",i,j);
    scanf("%d",&dist[i][j]);
   }
 floydWarshell();
 printf (" \n\n shortest distances between every pair
 of vertices \n");
 for (i = 0; i < n; i++)
 {
  for (j = 0; j < n; j++)
   printf ("%d\t", dist[i][j]);
  printf("\n");
 }
 return 0;
}
```

10.13 PROGRAM FOR THE KRUSKAL METHOD

```
#include<stdio.h>
#include<conio.h>
#include<stdlib.h>
int i,j,k,a,b,u,v,n,ne=1;
```

```
int min,mincost=0,cost[9][9],parent[9];
int find(int);
int uni(int,int);
void main()
{
 clrscr();
 printf("nntImplementation of Kruskal's algorithmnn");
 printf("nEnter the no. of verticesn");
 scanf("%d",&n);
 printf("nEnter the cost adjacency matrixn");
 for(i=1;i<=n;i++)
 {
  for(j=1;j<=n;j++)
  {
   scanf("%d",&cost[i][j]);
   if(cost[i][j]==0)
    cost[i][j]=999;
  }
 }
 printf("nThe edges of Minimum Cost Spanning Tree
 arenn");
 while(ne<n)
 {
  for(i=1,min=999;i<=n;i++)
  {
   for(j=1;j<=n;j++)
   {
    if(cost[i][j]<min)
    {
     min=cost[i][j];
     a=u=i;
     b=v=j;
    }
   }
  }
  u=find(u);
  v=find(v);
  if(uni(u,v))
  {
   printf("n%d edge (%d,%d) =%dn",ne++,a,b,min);
   mincost +=min;
```

```
   }
   cost[a][b]=cost[b][a]=999;
  }
 printf("ntMinimum cost = %dn",mincost);
 getch();
}
int find(int i)
{
 while(parent[i])
  i=parent[i];
 return i;
}
int uni(int i,int j)
{
 if(i!=j)
 {
  parent[j]=i;
  return 1;
 }
 return 0;
}
```

10.14 PROGRAM FOR DIJKSTRA METHOD

```
      #include<stdio.h>
#include<conio.h>
#define INFINITY 9999
#define MAX 10
void createGraph(int G[MAX][MAX],int n) ;
void dijkstra(int G[MAX][MAX],int n,int startnode);

int main()
{
      int G[MAX][MAX],i,j,n,u;
      char ch;
//input the number of vertices
      printf("Enter no. of vertices:");
      scanf("%d",&n);

      createGraph(G,n);
      printf("\nEnter the starting node:");
```

```
        fflush(stdin);
        scanf("%c",&ch);//read the starting vertex
        u= toupper(ch)-65;//convert to its equivalent
        numeric value i.e/ a=0,b=1,c=2 and so on....
        dijkstra(G,n,u);

        return 0;
}

//create the graph by using the concepts of adjacency
matrix.
 void createGraph(int G[MAX][MAX],int n)
 {
        int i,j;
 //read the adjacency matrix
        printf("\nEnter the adjacency matrix:\n");

        for(i=0;i<n;i++)
              for(j=0;j<n;j++)
                    scanf("%d",&G[i][j]);

  }

void dijkstra(int G[MAX][MAX],int n,int startnode)
{

        int cost[MAX][MAX],distance[MAX],pred[MAX];
        int visited[MAX],count,mindistance,nextnode,i,j;
        char ver='A';

        //pred[] stores the predecessor of each node
        //count gives the number of nodes seen so far
        //create the cost matrix

        for(i=0;i<n;i++)
              for(j=0;j<n;j++)
                    if(G[i][j]==0)
                          cost[i][j]=INFINITY;
                    else
                          cost[i][j]=G[i][j];

 //initialize pred[],distance[] and visited[]
```

```
for(i=0;i<n;i++)
{
            distance[i]=cost[startnode][i];
            pred[i]=startnode;
            visited[i]=0;
}

      distance[startnode]=0;
      visited[startnode]=1;
      count=1;

      while(count<n-1)
{
            mindistance=INFINITY;

            //nextnode gives the node with minimum
            distance
            for(i=0;i<n;i++)

                  if(distance[i]<mindistance&&!visite
                  d[i])
                  {
                        mindistance=distance[i];
                        nextnode=i;
                  }

      //check if a better path exists through nextnode
      or not
      visited[nextnode]=1;
      for(i=0;i<n;i++)
            if(!visited[i])

                     if(mindistance+cost[nextnode]
                     [i]<distance[i])
                     {
                         distance[i]=mindistance+cost
                         [nextnode][i];
                         pred[i]=nextnode;
                     }
      count++;
}
```

```
    //print the path and distance of each node from
    starting node
    for(i=0;i<n;i++)
          if(i!=startnode)
          {
          printf("\nDistance of node from %c to %c =
          %d",ver, ver+i,distance[i]);

                printf("\nPath=%c",ver+i);
                j=i;
                do
                {
                      j=pred[j];
                      printf("<-%c",ver+j);
                }while(j!=startnode);
    }
}
```

10.15 BFS USING COLOR CODE

```
/***********************************************************
 * Breadth First Search in Graph ( BFS ):                  *
 * Show minimum cost path from source to destination       *
 * using BFS at source .                                   *
 ***********************************************************/
#include<stdio.h>
#define MAX 100

#define WHITE 0
#define GREY 1
#define BLACK 2
#define EDGE 1

int i , j , m , n ;

int M , N ;

deque Q ;

deque :: iterator p ;
```

```
int graph[MAX ][MAX ];
int color[MAX ];
int dist[MAX];
int finish[MAX];
int parent[MAX ] ;

void printPath(int start , int end )

{
 if( start == end )
  printf("%d ",start);

 else if( parent[ end ] == -1 )
  printf("No path from %d to %d\n",start,end);

 else
 {
  printPath( start , parent[end ] );

  printf("-> %d ",end );
 }
}

void Bfs( int s )
{
 int u ;
      /** INITIALIZATION **/
 for(i = 1 ; i <= N ; i++)
 {
  color[ i ] = WHITE ;

  dist[i ]   = -1 ;
  parent[ i] = -1 ;
 }
 color[ s ] = GREY ;

 dist[s ] = 0 ;   /**INITIALIZE DISTANCE OF SOURCE TO
 ZERO**/

 parent[s ] = -1 ;

 Q.push_back(s) ;
```

```
while( !Q.empty() )          // LOOP UNTIL QUEUE(Q) IS
EMPTY
{
   u = Q.front();

   Q.pop_front();   /** POP FROM QUEUE **/

   for( i=1 ; i<= N; i++)
   {
 if( color[i ] == WHITE && graph[u][i] == 1 )
 {
  color[i ] = GREY ;

    dist[i ] = dist[u] + 1 ;

  parent[ i] = u ;

  Q.push_back( i ); /** PUSH INTO QUEUE **/
 }
   }
   color[u ] = BLACK ;
 }
}
int main(void)
{
// freopen("t.in","r",stdin);
// freopen("t.out","w",stdout);

 int a, b ;

 scanf("%d",&N);          /**Verticess (N) ******/
while( 1 )          /** INPUT EDGES ****/
 {
  scanf("%d %d",&a,&b);

  if( a== 0 && b==0 )
   break ;

  graph[a][b] = 1 ;
 }
```

```
 while( scanf("%d",&a) != EOF )        /* ENTER START
 AND END */
 {
  scanf("%d",&b);
           Bfs(a);    /** EXPLORE ALL NODE FROM START
           TO END **/
       /** BFS() CALLING WITH STARTING NODE **/

  printf(" Source: %d , Destination: %d\t",a,b);

  printPath(a,b);       /** printPath() CALL **/

  printf(" End of Path\n\n");
 }
 return 0 ;
}

// END OF CODE
```

Input:

```
---------
8
1 2
1 4
2 1
2 3
3 2
4 1
4 5
4 8
5 4
5 6
5 8
6 5
6 7
7 6
7 8
8 4
8 5
8 7
```

```
0 0
1 7
3 7
4 6
```

Sample Output:

```
--------------
 Source: 1 , Destination: 7, 1 -> 4 -> 8 -> 7  End of
Path

 Source: 3 , Destination: 7, 3 -> 2 -> 1 -> 4 -> 8 ->
7  End of Path

 Source: 4 , Destination: 6, 4 -> 5 -> 6  End of Path
```

Index

Printed in the United States
by Baker & Taylor Publisher Services